A DEO VICTORIA

The Story of the Georgian Bay Lumber Company
1871 - 1942

A DEO VICTORIA

The Story of the Georgian Bay Lumber Company
1871 - 1942

James T. Angus

Ontario Heritage Foundation
Local History Series #2

Severn Publications Limited
Orillia
1994

To free enterprise and all those who take risks
in pursuing goals to which they are dedicated.

Canadian Cataloguing in Publication Data

Angus, James T. (James Thomas), 1928-

  A Deo Victoria

(Ontario Heritage Foundation Local history series; no. 2)
Includes bibliographical references.
ISBN 0-9694197-0-8

1. Georgian Bay Lumber Company - History.
2. Lumbering - Ontario - Georgian Bay Region - History.
I. Title. II. Series.

HD9764.C34G46 1990 338.7'634982'0971315
C89-090597-5

Severn Publications Limited
95 Matchedash St. N. #404
Orillia, Ontario
Canada L3V 4T9
(705) 329-2127

This book was published with the assistance of the Ontario Heritage Foundation, Ontario Ministry of Culture and Communications.

Foreword

The Ontario Heritage Foundation Local History Series is a programme of financial and technical assistance in support of the publication of original works on the province of a local or regional nature.

The appearance of this second volume in the Local History Series, *A DEO VICTORIA: The Story of the Georgian Bay Lumber Company* by James T. Angus, is a testament to the maturing of the field of local history in Ontario. Professor Angus describes and analyzes the role played by the lumber industry in the development of many of the communities on the eastern side of Georgian Bay. His work reminds us of how the lumber business in the late nineteenth and early twentieth centuries shaped the growth of not only a locality but of entire regions in the province. More than just a first-rate business history, this book is a social history of the various people — from the lumber barons to the First Peoples of the Dokis Reserve — whose lives were tied to the harvesting of the Georgian Bay area forest. Good local history, Angus demonstrates, provides insights for many dimensions of the province's past.

The distribution by the Board of Directors of the Ontario Heritage Foundation of copies of this second volume from the Series will highlight, it is hoped, the integral contribution of local history to our understanding of Ontario's heritage.

Richard Alway, Chairman
Ontario Heritage Foundation

Preface

When I was a boy, growing up in the tiny community of Big Chute on the Severn River in the 1930s, the fabled Georgian Bay Lumber Co. was often spoken about by our elders. Even though the company had not logged the river for over forty years, there were plenty of relics around to confirm the loggers' one-time existence. Giant, fire-blackened pine stumps were scattered through the second-growth forest; iron rings, once used to anchor catch booms at the foot of the rapids, were still fastened to the shore; here and there, one could find initials chiselled by idle river drivers into flat rocks; water-logged sawlogs could be fished from the bottom of shallow bays; a square pine timber, still wedged in the rocks below the Big Chute dam, had defied generations of loggers to pry it loose; further back from the river, on tributary streams, one could find the rotting remains of log dams and timber slides.

Many of our older neighbours, at Port Severn and Waubaushene, had actually worked for the company in their youth – in the mills, in the forests, and on the river drives. Their stories were legion. To my brothers and me, these men were heroes. And there were the former officers and owners of the company, mainly summer residents – the Sheppards, Loverings, Buchanans, Stockings, Grays, and their descendants. One spoke to and about these people with deference, because they had what few of the rest of us possessed in the bleak 1930s: money – lots of money. They enjoyed all the privileges of wealth – fine homes, summer cottages, yachts, speedboats, and big cars. But for all that, they were ordinary folk, friendly and generous. Although the people and the relics were real, the company, which had sawed its last pine tree eight years before I was born, had become largely legend. Time passed. Memories faded.

Then a few years ago, while researching the history of the Trent-Severn Waterway, I was amazed by the frequency with which the Georgian Bay Lumber Co. and the names of some of its officers appeared in the data. Boyhood memories were rekindled; my interest in the company was revived. Even superficial research revealed that the origin of the company antedated by many years the generation of lumber-jacks and mill managers I knew in my youth and that there was more to its history than the romantic images that enriched my memory. Sensing that an interesting story lay shrouded by the mists of time, I decided to look deeper into the activities of this company.

First, I returned to my roots. I discovered that my community was gone. All the houses had disappeared, but, incredibly, some of the lumber workers I had known as a boy were still active and living nearby. There was Vic Conner, a close neighbour with whose children I had grown up. Vic had worked in the Waubaushene mill and in the lumber camps during the First World War. There was Billy St Amand whom I thought was already old when I was a boy. Billy was now ninety-four years old, but still alert and full of memories. He had worked in the Port Severn mill when he was a young man and was there when the mill burned in 1896. There was Clarence Russell whose, father had managed the company stores for over forty years. Clarence, too, had worked in the stores and had eventually taken them over.

Many others came forward and volunteered information, mostly reminiscences of their own or their parents. Nostalgia produced many myths and legends but few facts, so that I was soon able to resurrect the soul of this once great institution but had unearthed little of its corpus. Tedious digging later in archives, family records, contemporary newspapers, land registry offices, and recorded histories, gradually produced the facts out of which the skeleton of the company's story was reconstructed. What emerged is the anatomy of one of Ontario's nineteenth- century resource companies and the biographies of the men and women who created, re-created, and managed it for more than fifty years.

The Georgian Bay Lumber Co. was appropriately named. Unlike the dozens of other lumber companies, large and small, that harvested the virgin pine in the Georgian Bay district, but whose operations were in the main localized, the domain of this company was the entire Bay, from Collingwood to Blind River. Not only was it the largest lumber company of its time, in terms of area of operation and annual board-foot production, but it was the oldest, having been formed through the purchase and amalgamation of all the original Georgian Bay sawmills. Contemporary experts judged it one of the finest lumber companies in the province. It produced, through the years, hundreds of millions of board feet of prime-quality pine lumber, earned tens of millions of dollars for its handful of owners, and provided incomes for thousands of lumber workers and their families.

But the story is not just a record of the technical- commercial achievements of a collection of sawmills. Rather it attempts to chronicle the fortunes and misfortunes, successes and failures, of a small group of nineteenth-century entrepreneurs – American and Canadian – whose sole motive for exploiting the forests was the acquisition of wealth. Although the men and the companies they formed are the central focus, the story attempts to give some insight into the business strategies and ethics of the period.

The Georgian Bay Lumber Co. had its genesis in the entrepreneurial impulses that throbbed in early-nineteenth-century North America and produced the capitalistic system as we know it today. The Phelps and Dodge families of New York, founders of the Georgian Bay Lumber Co., were typical of the generation of American entrepreneurs whose exploits fostered the rise of modern capitalism. Their experience illustrates the traditional evolution of the system from commercial capitalism to industrial enterprise and finally financial investment. They started as merchants, then became traders in cotton and metals. They used their profits from these ventures for investment in the great growth industries of the nineteenth century – railways, real estate, banking, lumbering, and mining. The Georgian Bay Lumber Co. was an outgrowth of their extensive lumbering operations in the United States.

While the entrepreneurial ethic of the mid-nineteenth century gave birth to the company, its philosophy had its roots in the Calvinist-Puritan ethos of an earlier age. It was the marriage of these two traditions – entrepreneurial risk-taking and Calvinist zeal – that not only gave rise to the Georgian Bay Lumber Co. but sustained it through successive fluctuations in the business cycle.

Untold here are the exploits of the loggers whose tales of daring so enthralled me as a boy. Stories about them and their peers have been adequately recorded elsewhere. The subjects of this book are the captains of industry whose contributions to the Canadian lumber trade have been too often overshadowed by the more glamorous activities of the lumber-jacks who worked for them. What follows is the bosses' story.

Acknowledgments

This book has been published with the help of a grant from the Ontario Heritage Foundation. I am most grateful to the foundation for this support and to the foundation's staff members for their encouragement and assistance.

Many other organizations and individuals contributed to the production of the book. I gratefully acknowledge the invaluable assistance provided by the staffs of the following organizations: the Anglican Church Archives, in Toronto; the Archives of Ontario; the Collingwood Museum; the Connecticut Historical Society, Hartford; the Danville Public Library, Danville, Illinois; the Dokis Indian Reserve Band Office; the Huronia Museum, in Midland; Lakehead University Library; the National Archives of Canada; the National Archives and Records Service, Washington, DC; the Parry Sound Public Library; the Simcoe County Archives, at Minesing; the Vermilion County Courthouse, in Danville, Illinois; the Waubaushene Public Library; and, the West Parry Sound District Museum.

Numerous individuals kindly provided documents, photographs, and anecdotes, without which the story told here would be incomplete. I am especially indebted to Charles A. Stocking, Rissah Mundy, Jamie Hunter, Bob Thiffault, Bill Russell, Clarence Russell Sr, Billy St Amand, Frances Wellman, Emery Moreau, and Victor Conner.

Several of the grandchildren of W.J. Sheppard deserve my special gratitude. I should like to thank in particular Margaret Bell, Gerald Gray, Gladys Kelly, Eleanor Sheppard, Reg Sheppard, Robert Sheppard, and Nora Vaughan, whose memories assisted me in reconstructing the parts played in the story by their parents, grandparents, and other relatives.

I owe a debt of gratitude to Mrs Cleveland E. Dodge Jr, who kindly gave me copies of letters by and about members of the Phelps and Dodge families and for providing other information about these families not available in published sources. I am grateful, also, to Mrs Pauline Webel, for allowing me to read portions of the then-unpublished manuscript of her biography of her grandmother, Josephine Dodge.

I acknowledge with appreciation the permission of the Ontario Historical Society to include in this work material on the Dokis Indians published previously in Ontario History (September 1989).

Finally, I express my thanks to my copy-editor, John Parry, who ensured for a second time that my writing was both textually consistent and grammatically correct.

Part One:

First Empire: The Rise and Fall of Anson Dodge (1867-1873)

Anson Green Phelps Dodge, 1834-1918

Chapter One

"Alphabet" Dodge

The founder of the Georgian Bay Lumber Co. was Anson Greene Phelps Dodge. His parents called him Anson, but Canadian lumbermen knew him as Alphabet Dodge because of his habit of signing his name with his initials — A.G.P. Dodge.

An American citizen who came to Canada in 1866 to participate in the lucrative lumber trade, A.G.P. Dodge became a highly controversial Canadian citizen, enigmatic in the extreme — saint or sinner, depending on which contemporary newspaper one chooses to believe. The Bracebridge *Northern Advocate* described him as a man "with high moral training and advanced mental culture,"[1] while the Toronto *Globe* called him a liar, an example of a "Tweed" or "Connolly," a traitor to his homeland, and an embarrassment to his friends, and the paper accused him of writing with poor grammar and incorrect spelling. The Toronto *Mail* judged him "a truly liberal and progressive man,"[2] and the Newmarket *Era*, though not always friendly, allowed that Dodge was "upright in his deportment and a very worthy citizen."[3] The Orillia *Northern Light* was the most flattering of all: "no man [was] more popular in this country nor [had] done more to advance its interests"[4] than A.G.P. Dodge.

Clearly, Anson Dodge had a knack of arousing genuine affection or open hostility in those who knew him during his brief sojourn in Canada. Even his name represented a curious mixture of loyalties and allegiances. He bore the name of his maternal grandfather, Anson Greene Phelps, a master saddler from Hartford, Connecticut, who would amass a fortune in the drygoods business and would eventually, in partnership with his son-in-law, William Earl

Dodge (A.G.P.'s father), found Phelps Dodge and Co., now Phelps Dodge Corp.

Grandfather Anson Greene Phelps, born in 1781, was christened Anson in honour of the renowned English admiral, George Anson, who circumnavigated the globe between 1740 and 1744, harassing Spanish shipping, and later became First Lord of the Admiralty. The name Greene testified to his father's admiration for Gen. Nathaniel Greene, the commanding officer under whom he served in the American Revolutionary War. By word and deed, the namesake would demonstrate, before he left Canada, a similar ambivalence toward the institutions of Britain and those of his native United States.

Anson seems to have been a mimic, adopting the manners, dress, and habits of those whose successes he wanted to emulate. Fred Cumberland, managing director of the Northern Railroad, of whom it was said Anson was a protege, was an impeccable dresser; Anson dressed impeccably, too, patronizing the most fashionable tailor shops in Toronto. Anson's younger brother, David Stuart, and his own son, Anson Greene Phelps Dodge Jr, were, respectively, Presbyterian and Episcopalian ministers. In later years, when fame and fortune had slipped through his own fingers, Anson mimicked his brother and son, often posing as a clergyman, dressing in a frock coat and wearing a stringy little black tie. Another brother, Charles C. Dodge, was a general in the US Army. Anson became a "general," assuming the rank when he went into semi-retirement. He would be buried as Gen. Dodge, none of his acquaintances, perhaps not even his wife, apparently knowing that the military distinction was self-styled.

Paradoxically, the man whom he imitated most consistently was his father, the loving patriarch, his own dependence on whom Anson seemed to resent and against which he struggled most of his life. His father had made a great deal of money in the lumber trade; Anson too sought wealth in lumber. His father was a railway tycoon; Anson invested in Canadian railways. His father was a US Congressman; Anson became a member of the Canadian Parliament. His father owned a magnificent summer home at Tarrytown, New York; although he could ill afford one, Anson built a mansion on the south shore of Lake Simcoe, on land that would later become valuable summer property for the well-to-do of Toronto but, at the time, was little more than marginal farm land, recently denuded of stands of pine timber.

His father was a renowned philanthropist; Anson once tried to engage in philanthropy but, as with all his other ventures, the effort failed, his commitment to charity exceeding his ability to deliver. At the end of the Civil War, while just starting out in business, he learned that the Theological Seminary and High School of Alexandria, Virginia was having financial trouble and about to close. On an impulse, "without consulting anybody, and without knowing how poor [he] was or whether [he] had the means or not,"[5] he advanced the school $100,000 in promissory notes on which he would pay the annual interest, a promise he later had difficulty keeping.

A journalist for the Hamilton *Times* left us a not very flattering description of A.G.P. Dodge, written in the spring of 1873, while Anson was sitting in Parliament. "Mr Dodge looks 37 or 38 years (his actual age was 39), has sandy and curly hair, parted with great care. His wiskers etc. are also light. He has not a bad looking face, but there is little character in it . . . He has the New York swell accent in its purity, and postures and gesticulates in the most approved American style."[6] Most contemporary correspondents agreed with the *Times* writer that Anson had "not distinguished himself in Parliament."[7]

Whatever qualities Anson may or may not have had, he had one attribute on which all his critics were agreed: he was a man with boundless energy and insatiable ambition. The always friendly *Northern Advocate* said: "His movements are the most rapid upon record and his punctuality so well known that he is spoken of as a model."[8] The *Mail* described him as "a fine public spirited man, of large views and wonderful enterprize."[9] Even the *Globe* had to admit that he was "an active speculator, fond of anything exciting, and what is called a jolly fellow."[10]

A.G.P. Dodge was no ordinary immigrant. For one thing, he was rich; at least, he acted as though he were. When he came to Canada, the assets of his firm, Dodge and Co., were estimated to be worth from $600,000 to $800,000. Anson claimed that his father estimated his (Anson's) net worth at $750,000, while his older brother, William, whom Anson considered "the most conservative man in New York,"[11] estimated the value of Anson's company at $600,000. Compensating for his brother's conservatism, Anson added another $200,000 and generally let it be known that he was worth $800,000.

But most of Anson's money was tied up in timber lands and logs in Pennsylvania and lumber in his yard in Philadelphia. He had little fluid capital to invest. Anson's father, the Hon. William Earl Dodge, was, however, an extremely wealthy man. Being one of New York's leading citizens, with international connections, William Earl was well known in Canadian business and government circles. And so, with his own alleged wealth, and backed by his father's reputation, Anson had no difficulty finding creditors to finance the wild spending spree in which he engaged between 1869 and 1872. In less than three years, he bought, built, and renovated no fewer than eight sawmills around the shore of Georgian Bay. He acquired steam tugs, barges, horses and oxen, harness, sleighs, wagons, chains, tools, shanties, and shanty furniture and equipment — all the paraphernalia needed to carry on a giant lumber business. He built homes and boarding-houses for his employees. He acquired hundreds of square miles of timber limits. He invested in two railways and a grist mill and, as was stated, built a magnificent mansion on Lake Simcoe. In a couple of years he became one of the largest lumber and timber manufacturers in Ontario, perhaps second in size only to the fabled J.R. Booth, who lumbered in the Ottawa Valley. Had his father, who eventually bore the financial burden of Anson's reckless spending, known what his son was doing in the wild woods of northern Canada, he undoubtedly would have counselled greater caution.

The second son in a family of seven boys (two sisters died in infancy), A.G.P. Dodge was born in New York, on 24 August 1834. Like all his brothers, he was given a liberal education and training and all the assistance needed to establish a career. Members of a close-knit family with strong religious convictions and conservative tendencies, six of the seven sons seemed content to pursue the careers mapped out for them by their father; consequently all of them achieved a high degree of success, whether in a business or a profession. Anson was the exception. While not averse to accepting financial assistance from his father, he chose to follow his own course. In temperament, he was fundamentally different from his brothers. While they were cautious and conscientious, Anson was restless, venturesome, and a risk-taker. Nor was he nearly as successful as they.

After finishing school, Anson travelled widely through Europe. From Europe he went to Australia, where he lived for two years. Then in 1855, at the age of twenty-one, when he thought himself ready to embark on his first business venture, he set up a co-partnership with his father's assistance — Bussing, Crocker and Dodge. The partners operated out of the head office of Phelps Dodge and Co. on Cliff Street, New York, selling, on commission, the products of four or five of the subsidiary companies of Phelps Dodge. Sales were made in the New England states.

After a few years, Anson grew tired of selling and, seeking adventure, moved to Marquette, Michigan. Completion of the Sault Ste Marie locks in 1855, and construction of an ore dock at Marquette in 1857, made possible the development of vast ore deposits in northern Michigan. Anson managed a blast furnace in Marquette and supervised mining operations for one of the mining companies. But he soon became tired of mining and decided to enter the lumber business.

One of the principal subsidiary enterprises of Phelps Dodge was the lumber business in which the company had engaged since 1835. To provide an outlet for its lumber products, the firm had established a large lumber yard in Baltimore under the firm name of Henry James and Co., Henry James being the resident partner and general manager of the operation. On Anson's behalf, William Earl wrote to his brother-in-law and partner Daniel James in the spring of 1859 to inquire about the possiblity of Anson joining the Baltimore firm to learn the trade. Daniel was somewhat reluctant to have Anson join the firm. He wrote guardedly to William Earl:

> I note what you say about your son Anson and his abandoning the iron concern. I think it is a favourable move for him. It is most desirable for us to have someone in Baltimore to assist Henry . . . . It will of course take him some time to learn the business . . . especially as it is so entirely different from anything he has ever been accustomed to. He has had so many changes I fear he would go all right for a time and then want to try something else . . . If he would give his mind to it and go at it and adhere to it as his business for life, then I should say it would be most desirable for us and him . . . If he was married I would have no fear of him.[12]

Whether he fell genuinely in love, or whether he just wanted to ensure a position with Henry James and Co., Anson quickly found the wife Uncle Daniel required as a condition of employment. On 12 October 1859, he married Rebecca Wainwright Grew, a nineteen-year-old from Alexandria, Virginia. Anson and Rebecca settled in Baltimore, Anson to learn the lumber trade and Rebecca to start raising a family, for the following year their son, Anson Greene Phelps Dodge Jr, was born.

Typically, Anson grew restless in his uncle's firm in Baltimore and decided to break out on his own. In May 1864, he established his own lumber company — Dodge and Co. — at Williamsport, Pennsylvania. He had two junior partners, Titus Benjamin Meigs, a specialist in lumber operations, and Charles Hebard, an accountant. In 1867, Anson's young brother, Norman White, joined the firm. Anson controlled 80 per cent of the company, which he launched with $30,000 inherited from his grandfather and some capital he borrowed from his father. With this he bought 20,000 acres of pine lands from his grandfather's holdings at Tobyhanna in Monroe County, Pennsylvania. As the firm prospered, he bought other pine lands, cut off the logs, and ran them to Williamsport, where his father allowed him to saw the logs in his mill.

The Dodge mill at Williamsport, built in 1864 at a cost of about $100,000, had a cutting capacity of 50 million board feet per year and was reputed to be the largest enterprise of its kind in the world. As an outlet for the 20 to 25 million board feet of lumber annually produced, Dodge and Co. bought a lumber yard in New Jersey for $90,000, half of which money was put up by Anson's father. During the Civil War and the period of reconstruction that followed, Dodge and Co. prospered, as the price of lumber reached an all-time high of $20 per thousand feet, board measure. But despite the firm's success, or maybe because of it, the partnership broke up in 1869.

Anson had begun speculating in commercial ventures that were not connected with the business. It was during this period, for example, that he bought his first timber rights in Canada. The more conservative Meigs and Hebard and brother Norman, alarmed at Anson's investment recklessness, withdrew from the company and, along with George Egleston, the next Dodge brother in succession, formed a new firm, Dodge and Meigs. The brothers took over the New Jersey lumber yard and half of the Tobyhanna property.

With what was left, Anson formed a new Dodge and Co. with partners William Jay Hunt and Samuel Schofield. Formerly clerks in the earlier company, Hunt and Schofield were each given one share in the new company, which guaranteed them an income of $1,500 to $2,000 a year. Hunt was the bookkeeper, and Schofield and his brother were placed in charge of a new lumber yard in

Philadelphia. The head office of the company was in New York. From then on, Anson had no direct business connection with any of his brothers.

Anson Dodge first learned about the possibility of lumber speculation in Canada from Henry W. Sage of Brooklyn, New York. Known in the American lumber fraternity as "King of the Trade," Sage was one of the owners of the largest holdings of pine property in the United States and was recognized as one of the country's most successful lumbermen. He had built a lumber mill in Canada West (now Ontario) at Belle Ewart on the south shore of Lake Simcoe in 1852, one year before the Northern Railway reached Belle Ewart and two years before the Reciprocity Treaty of 1854-65 stimulated the Canadian lumber trade by opening tariff-free American markets to Canadian lumber. Sage, who operated the Belle Ewart mill under the name of Sage and McGraw (later Sage and Grant), told Anson that he had made a personal profit of $900,000 in a very short time from the operation.

Anson became seriously interested in the pine lands in Canada in 1866, when he learned from Sage that large tracts of pine in Parry Sound and Muskoka districts would soon be available for sale by the Department of Crown Lands of the United Province of Canada and that more would be sold in the future. In fact, Sage had bought rights to sixty-five square miles of timber in Oakley township, near Bracebridge, the supply of pine in the southern half of Simcoe County, in the area surrounding Belle Ewart, having already been consumed by the myriad lumber mills that sprang up in the wake of the Northern Railway. Sage showed Anson letters from the commissioner of crown lands and others interested in the lumber trade in Canada, indicating "the most wonderful exhibit of possibilities in connection with Canada timber lands that anyone interested in pine timber lands could desire."[13] Sage told Anson that if he were twenty years younger he would sell all the land he owned elsewhere and invest in Georgian Bay pine lands, because he considered that no comparable property had been put on the North American market for years. Anson came to Canada to examine the timber lands for himself. Sage wrote him letters of introduction to the commissioner of crown lands and other politicians and lumbermen in Canada with whom he had done business. In the summer of 1866, Anson arrived in Toronto and then travelled north to Newmarket, Belle Ewart, Barrie, and Orillia, where the northern lumber trade was then centred.

Political changes were taking place in Canada that Anson only vaguely understood, but he sensed a mood of optimism and excitement in business circles, as final arrangements for the new Confederation of Canadian provinces were being worked out in Ottawa and London. Those who would participate in the new Ontario government, to be based in Toronto, were already talking about immigration policies to increase the population and thus the economic base of the province. The pine lands in Muskoka and Parry Sound districts would be set aside as "free grant" land for immigrants, but, first, timber rights to this huge block of land — ninety miles long and seventy-five miles wide — lying between the French and Severn rivers would be sold to lumbermen at public auction. The Crown Lands Department estimated that this area contained about one-eighth of all the standing pine timber remaining in eastern Canada. Opportunities for investment were unlimited, and Anson decided to get in at the beginning.

After discussing purchase procedures, fees, and timber regulations with officials in government, and after talking about logging operations and markets with lumbermen, Anson set out to examine the pine lands in the north. There were two principal routes into Muskoka and Parry Sound districts in 1866. One could take the train to Collingwood, the northern terminus of the Northern Railway, and thence travel by steamer up the northeast shore of Georgian Bay, disembarking at the mouths of any of the several rivers that drained the hinterland. Or one could take the inland route by way of Washago, the Muskoka Road to Gravenhurst, and thence by canoe through the network of lakes and rivers that covered the Canadian Shield. Anson chose the latter route.

From Orillia, he took the regular ferry to Washago Mills; from Washago, he travelled by stage-coach fourteen miles on the Muskoka Road to the developing village of Gravenhurst, situated on a gentle declevity between Muskoka and Gull lakes. Gravenhurst had one hotel — the Freemason's Arms — operated by T.B. Horton and a general store owned by A.P. Cockburn, member of Parliament for Muskoka and Parry Sound. An Anglican church, which was under construction, and a handful of settlers' homes rounded out the village. Nearby was a lumber camp operated by A.P. Cockburn, who had commenced lumbering operations the previous winter. A quarter-mile to the west, on the shore of Muskoka Bay, was a

400-foot long wharf for Cockburn's sixty-two-ton steamer *Wenonah*, launched in June and put into regular service just days before Anson's arrival.

From Gravenhurst, Anson travelled up Lake Muskoka, probably on Cockburn's steamer, to the Indian village of Port Carling. At Port Carling, he crossed the portage into Lake Rosseau; travelling by canoe, he reached the village of Rosseau at the north end of the lake. From Rosseau, his journey took him by wagon along the recently opened road to Parry Sound, where John and William Beatty operated a small sawmill.

Anson's destination was Byng Inlet, some forty miles north of Parry Sound, where he had been advised a mill site was available for lease at the mouth of the Magnetawan River. Having reached Byng Inlet, he undoubtedly ascended the sluggish-flowing river to its headwaters to evaluate the quantity of pine available to supply a future mill. From the Magnetawan, he worked his way back through the lakes and rivers to Gravenhurst and eventually back to Belle Ewart and home.

Just as his father had explored the remote regions of Pennsylvania on horseback thirty years earlier, before investing in timber, so Anson had travelled by canoe through the wilder and even more remote vastness of the Georgian Bay hinterland. Had he been an ordinary tourist, seeking adventure and pleasure, Anson would have been awed by the primeval magnificence of the virgin forest — stands of red and white pine, some trees five feet in diameter — that covered the unmolested landscape.

But A.G.P. Dodge was a merchant, with aspirations of becoming the largest producer of lumber in North America. Entrepreneurial greed allowed him to see the huge conifers only in terms of investment. Screening out the natural beauty of the forests, his calculating eyes measured only the potential of the trees for generating wealth. Each large tree contained "x" hundreds of feet of lumber, each square mile of forest contained "y" number of trees. These virgin forests contained hundreds of millions, if not billions, of board feet of prime-quality pine lumber, and each thousand feet would fetch a profit of from four to five dollars on the New York market. The enormity of potential profits was staggering.

Ten years later, Anson described the dreams that danced in his head as he meditated around his camp-fires on that northern pilgrimage. His dreams were an admixture of a young man's vision, a nineteenth-century morality of getting rich through the indiscriminate destruction of natural resources, and an inherited Dodge obsession with accumulating great wealth. "My impression then was in estimating the vast magnitude of the transaction and the amount of capital to carry it on, that my best course was to sell out my timber lands in Pennsylvania which had run up a very high price — had reached, as I thought their maximum — and buy these government lands at what seemed to me the minimum price,"[14] Anson said later. "My own judgment, and my own impression was that whoever held ⅛ of the standing timber in Canada, and could successfully carry it through, would be one of the wealthiest men on the continent."[15]

And there is no question that Anson wanted to be that person. If for no other reason, he wanted to prove — to his father and to his uncles, all of whom thought him a bit of a profligate, and particularly to his older brother William, in whose shadow he always walked — that he, too, had what it took to become successful.

"My father's people, my grandfather — my family — had made millions of money out of similar transactions, and people all around me on every side whom I had acquaintance with, had been successful in such enterprises. My hope and my desire, was to convert all the property that I had into the form of timber property, and while free to admit that it soon took the form of insanity, I went into it with the best judgment that I had at that time and with the expectation of coming through and being wealthy from being a holder of large bodies of pine forests,"[16] Anson later recalled. But alas! as this story will reveal, it required more than dreams to become wealthy in the competitive world of the nineteenth century.

Chapter Two

Anson's Insanity

After his exploratory journey into Muskoka and Parry Sound districts, during which he committed himself to investing in Canadian pine, Anson Dodge began buying mill properties and acquiring timber licences with all the fervour of a committed Monopoly player. The first property he bought was in Byng Inlet. On 10 May 1867, Alex Macaulay, a lumberman from Orillia acting as an agent for Dodge, obtained licences for 364 square miles of timber lands, on the upper Magnetawan River, in the unsurveyed townships of Croft, Ferrie, Ryerson, Spence, and Wilson, and a lease of the water-power at the rapids at the mouth of the river. Two weeks later, Macaulay transferred the licences and the lease to Anson.

The fact that Anson acquired a water lease suggests that he may have contemplated building a water-powered sawmill in Byng Inlet; if so, he changed his mind and constructed a steam-powered mill instead. For this purpose, he applied for crown patents for 148 acres of land on the north side of the inlet, just below the rapids, and a larger block of 435 acres, covering both sides of the inlet, between the mill site and the Still River. These sites were surveyed by Thomas Bolger, PLS, and registered in the fledgling Ontario Crown Lands Department on 7 July 1867. In 1869, Anson acquired another 559-acre block for a mill village on the site of the present village of Britt.

It is not clear when Dodge's mill — the "Anson Mill" — was built. What is known is that Anson was carrying on extensive lumber operations on the Magnetawan in 1869, as reported by land surveyor Vernon Wadsworth, who visited the lumber camps in the winter. Also in 1869 Anson took up residence in Canada. The Anson

Map by Iain Hastie

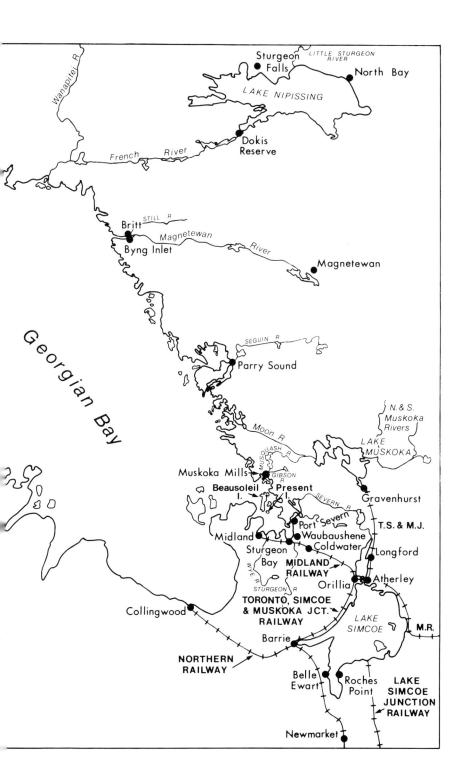

Sturgeon Falls

LITTLE STURGEON RIVER

North Bay

LAKE NIPISSING

Wanapitei R.

French River

Dokis Reserve

Britt

STILL R.

Magnetewan River

Byng Inlet

Magnetewan

Georgian Bay

SEGUIN R.

Parry Sound

N. & S. Muskoka Rivers

LAKE MUSKOKA

Moon R.

MUSQUASH R.

Muskoka Mills

GIBSON R.

Beausoleil I.

Present I.

SEVERN R.

Gravenhurst

Port Severn

T.S. & M.J.

Midland

Waubaushene

Coldwater

Longford

Sturgeon Bay

WYE R.

MIDLAND RAILWAY

Orillia

Atherley

STURGEON R.

TORONTO, SIMCOE & MUSKOKA JCT. RAILWAY

Collingwood

LAKE SIMCOE

M.R.

Barrie

NORTHERN RAILWAY

Belle Ewart

Roches Point

LAKE SIMCOE JUNCTION RAILWAY

Newmarket

Mill was originally owned by Dodge and Co.; under that name the lumbering operations were first conducted. Even though the Anson Mill ceased to be a branch of Dodge and Co., and even though the parent company ceased to exist after 1873, the name still lingers in the lumbering folklore of Parry Sound district.

Erecting a sawmill on Byng Inlet was not an easy task in 1867. Because no one lived within miles of the place, all the construction workers had to be imported. Bricks for the chimney, the steam plant and boiler, the milling equipment, wagons, sleighs, horses, oxen, and all the other paraphernalia used in the lumber business had to be taken to the mill site by steamer and barges from the railhead at Collingwood. Anson built, in addition to the mill, a blacksmith shop, carpenter shop, ice-house, two barns, a store, an office, freight house, steamboat dock, a manager's house, a millman's house, and twelve dwellings for workers. There were piling grounds, loading wharfs, and two farms, one on the Still River, the other at the rapids. The total cost of the real estate was $130,593.89.[1] It was reported in 1871 that the company was producing 25 million feet of squared timber and sawn lumber annually.[2]

Because the timber limits were located nearly forty miles upstream from the mill, freight destined to the camps was hauled by team along a winter "cadge" road to a depot at Port Anson on Ahmic Lake, situated four miles from the present-day village of Magnetawan. The road cost about $5,000 to build. The depot camp and a nearby farm, on which cows, horses, and oxen were kept, cost another $5,000, and another half-way camp for emergency use was built for $100. The horses, oxen, sleighs, and personal property for the surrounding camps were valued at $35,500 in 1871.[3] From the depot, food for the men and hay and oats for the horses and other supplies and equipment were distributed to the logging camps.

The winter's production of saw logs and square timber was floated on the spring freshet down the Magnetawan River to the inlet. Sawn lumber was shipped by schooner to Cleveland, Tonawanda, Albany, and Chicago. Squared timber was towed by steamer to Collingwood, loaded on the Northern Railway, hauled to Belleville, and rafted to Montreal and Quebec for shipment to England.

Creation of such an enterprise required a great deal of capital, but, more important, it demanded considerable energy and daring, both of which Anson possessed in liberal quantities. Investing in one remote lumber mill would have offered challenge enough for most men, but for Anson, an inveterate gambler, the Byng Inlet mill was only a beginning. Within four years, he would be in possession of seven more.

With the physical organization in place, Anson then set about creating a corresponding corporate structure to legitimize his Canadian operation. Never one to operate on a small scale, he employed one of the leading lawyers in the country — D'Alton McCarthy, an Irish-born resident of Barrie, Ontario. McCarthy was not only well versed in corporate and business law, but he had close political connections with the Conservative governments of John A. Macdonald in Ottawa and Sandfield Macdonald in Toronto.

Because of the international nature of Anson's lumber business, McCarthy advised securing a federal charter for the company. Accordingly, in February 1871, Anson, his minor partners, and a group of Canadian politicians applied, under the Canada Joint Stock Companies Letters Patent Act, 1869, for a charter incorporating the Maganettewan[4] Lumber Co. Letters patent[5] incorporating the firm were issued on 1 April 1871. The company was empowered to carry on the lumber business in all its branches and to establish a line of steamboats to run between Byng Inlet and Chicago and Cleveland for the purpose of carrying lumber and other freight. Capital was set at $250,000; each share was valued at $100. Since Dodge and Co. held the majority of the shares, Anson was president. C.H. Dill was mill manager, W.D. Kintzing was secretary-treasurer, Levi Miller was manager of logs and woods, and Capt. Isaac May was manager of vessels and transportation.

Meanwhile, another American company had built a sawmill on an island on the south side of the Inlet, about one mile from the Anson Mill. In 1866, Eli Clinton Clarke, a lumberman from New York and later Toronto, had obtained a timber licence for 192 square miles in Brown, Mowat, and Wallbridge townships. Clarke later teamed up with the firm of White and Co., which had headquarters in New York. The partners in this firm, Samuel White Barnard, Alanson Sumner Page, and Douglas L. White, although operating in Canada under the name of White and Co., had their own individual companies in the United States. These were combined into a larger firm — the International Lumber Co. — which had large lumber outlets in Albany, Chicago, and Oswego. The company had three lumber mills: one in Oswego and two in Canada — the Eagle and Star mills on the Bay of Quinte, near Belleville.

The Anson Mill – Byng Inlet c1880   – Ontario Archives 58256

The Page Mill – Byng Inlet c1880   – courtesy C.A. Stocking

In 1869, Clarke applied for a crown patent for 400 acres on the site of the present village of Byng Inlet. Here Clarke, White and Co. erected its sawmill, known as the Page mill, and established a mill village similar to Anson's on the opposite shore. In the same month that Anson applied for a federal charter, Clarke, White and Co. was granted an Ontario charter by provincial statute[6] for a company called the Georgian Bay Lumber Association,[7] with capital stock of $300,000. The company was empowered to buy the Page, Eagle, and Star mills, which the partners surrendered in exchange for stock.

In 1872, Anson and Clarke, White and Co. decided to join forces, a union that Anson welcomed, because it would facilitate the disposal of the 80 to 100 million board feet of lumber he hoped to produce annually in the several mills he was acquiring on Georgian Bay. One of the reasons for the large overhead costs in the manufacture and handling of lumber was the amount consumed by "middlemen," who normally skimmed off 10 to 12 per cent of the selling price. An additional burden in the manufacture of lumber, peculiar to the remote Georgian Bay area, was the large cost of freighting the lumber or square timber to American and British markets. Further, Canadian lumber entering the US market was subject to an import tariff of $2 per 1,000 board feet.

To help reduce overhead costs, Anson had formed partnerships with small distributing firms in the United States: Barton and Spenser of New York; Chambers and Co. of Cleveland; and Perry and Packard, of Elizabeth, New Jersey. This arrangement was not entirely satisfactory, because the small firms were not able to handle all the lumber produced; moreover, Anson had to pay the high freight costs to the outlets. He was losing money and experiencing difficulty in paying interest and meeting mortgage payments. What he needed was an association with one large organization that could handle all his production and share some of the transportation costs. This he found in White and Co.

An amalgamation with Dodge would be advantageous as well to the partners in the International Lumber Co., especially White and Co. The latter had a large lumber outlet in Albany, then the largest eastern market for lumber. Selling three or four times as much lumber as any other dealer, White and Co. was by far the largest dealer in Albany. The company had, along with outlets in Oswego and Chicago, facilities for handling all the lumber Anson could produce in Canada. The International Lumber Co. wanted to obtain access to Anson's huge lumber production, at a fair rate, because the pine in Hastings County that supplied the Star and Eagle mills was rapidly running out.

The partners agreed to pay Anson $8 per thousand board feet — the estimated cost of getting the lumber out — for all his production; they would pay freight costs and dispose of the lumber at a rate of commission not to exceed 4 per cent of the selling price. After deducting all the costs, White and Co. would pay Anson one-half the net profit. In this way, Anson hoped to realize a profit of two to three dollars per thousand feet, which would allow him to meet mortgage and interest payments.

Because the two Byng Inlet mills controlled all the timber rights on the Magnetawan River, and because the companies were, to some degree, working at some disadvantage to each other, they decided to unite formally. Anson was allowed $200,000 for the Anson Mill, in return for which he received 25 per cent interest in the International Lumber Co., and $30,000 in cash. The partners then applied for a provincial charter for the Magnetawan River properties. On 2 March 1872, the Maganettewan Lumber Co. of Ontario, with share capital of $700,000, was incorporated by provincial statute.[8]

In 1856, the Gibson brothers, J.A. and William, sons of the Hon. David Gibson of Willowdale, Ontario, acquired timber licences for 100 square miles surrounding Parry Sound, on land that would later become the townships of Carling, Conger, Cowper, Foley, Gibson, and McDougall. Gibson[9] on whose behalf his sons had taken out the licence, built a water-powered sawmill, in 1857, at the first rapids above the mouth of the Seguin River. In the fall of 1863, the Beatty brothers — John and William — from Thorold, Ontario, bought the mill and timber rights from Gibson. The Beatties extended the timber holdings to 234 square miles.

In the early 1870s, the Rathbun brothers, who operated sawmills and other enterprises at Mill Point on the Bay of Quinte, concerned about the depletion of pine in Hastings County, decided to invest in Georgian Bay timber land. In December 1871, they bought the Parry Sound mill and timber limits from J.H. Beatty and Co. for about $75,000. They also bought an additional twenty-nine square miles of timber limits in Ferguson and McKellar townships in the crown timber auction of 1871.

The acquisitive Anson decided to add the Parry Sound property to his growing lumber empire, and the Rathbuns were willing, for a good price, to sell. Accordingly, on 1 June 1872, they sold Anson the mill, mill dam, wharfs, store-houses, timber limits, all the logs and lumber on hand, and a small steam tuge, *Wave*, for $154,064. By the terms of the indenture, Anson agreed to pay the Rathbuns $10,000 in cash, a further $49,064 in quarterly instalments of $10,000 at 7 per cent interest, and a final block payment of $20,000 in gold within one year of the sale. The Rathbuns gave Anson a mortgage of $70,000 for the balance. In return, the Rathbuns agreed to discharge the $75,000 mortgage that the Beatties had given them.

In March 1872, Anson incorporated, by provincial statute,[10] the Parry Sound Lumber Co., capitalized at $300,000. Anson held two-thirds of the shares. In the years ahead, the company would become a profitable enterprise, but when Anson bought the mill, it was quite small, employing only twenty-nine hands; it had a cutting capacity of about 20,000 board feet per day. The cost of production was low, averaging about $1.34 per thousand board feet for sawing, trimming, and piling.

Not only were production costs low, but the cost of logging operations was minimal too, none of the standing timber being located more than five or six miles from the mill. The logs were floated down the Seguin River or run down streams into Georgian Bay and towed a short distance to the mill. Parry Sound harbour, the best on the northeast shore of Georgian Bay, provided excellent accommodation for schooners. John C. Miller, a shareholder and future owner of the company, was mill manager and superintendent. Anson immediately put a gang of men to work enlarging the mill.

Muskoka Mills is located at the mouth of the Musquash River, a southern branch of the Muskoka River which drains the central region of Muskoka district. The first sawmill was built at Muskoka Mills in 1853 by Penetang merchant William Hamilton.

Several economic forces acting simultaneously in the early 1850s created strong markets for Canadian lumber and were, therefore, responsible for exploitation of the Georgian Bay pine forests and indirectly for construction of the first of the sawmills erected at Muskoka Mills. The Reciprocity Treaty permitted most Canadian natural products and raw materials, including sawn lumber, to enter American markets free of duty. The outbreak of the Crimean

War (1853-56) restored British markets for Canadian square timber and agricultural products which had been considerably restricted after repeal of the Corn Laws in 1846. Money made available to railway contractors by the Guarantee Act of 1849 and the Municipal Loan Fund Act of 1852 created the first Canadian railway building boom. The Free Grant and Homestead Act of 1853 encouraged immigration. The needs of communities that sprang up along the railway lines gave a further fillip to the lumber industry, and, with more money in the hands of farmers and merchants, secondary demand was created. Start of construction in 1853 of Ontario's first rail line — the Northern Railway — which, when completed, would open up Simcoe County and provide a direct link between Georgian Bay and Toronto produced a frenzy of sawmill construction in Simcoe County and around the shore of Georgian Bay. William Hamilton, who later admitted to having "caught the infection,"[11] obtained a timber licence for seventy-five square miles of forest on the Musquash and Gibson rivers and invested all his savings in a small, water-powered mill which he built at the first falls above the mouth of the Musquash. He hoped to ship his lumber to Chicago by schooner, but, by the time the mill began producing, peace had come to the Crimea and was followed by the inevitable depression; consequently the first hectic speculative phase of construction in Canada came to an end. Markets for Canadian lumber, both domestic and international, collapsed. Hamilton had difficulty disposing of any lumber. He reported that a shipload of two-inch pine planks shipped to Owen Sound for sidewalks "rotted in the pile,"[12] and he was never paid for it.

In 1857, Hamilton sold the Muskoka mill and timber limits to Charles Kelly, of Hamilton, for $16,000. Kelly was backed by lumber merchants in Chicago and Buffalo. Despite Hamilton's advice not to make any substantial improvements on the property until the price of lumber rose, Kelly spent considerable capital enlarging the mill. Then, as so often in Canadian history, crop failure coincided with economic recession, serving to deepen the depression of 1857-58, and Kelly was forced out of business. Hamilton lost the $12,000 mortgage he held on the mill.

The mill was next taken over by J. Tyson of Collingwood, but he also experienced financial difficulty. In 1861 the mill was reported to be idle.

In the mid-1860s, the American market for Canadian lumber began to grow again, as the population in the eastern United States increased rapidly. Settlement was also pushing west onto the prairies. Consequently, huge quantities of lumber were needed for building purposes in New York and Chicago. Since the Pennsylvania forests were practically exhausted and the pine in Michigan was also disappearing rapidly, American lumbermen looked north to Canada for a supply of sawlogs. Some lumbermen even built mills in Canada. Two Americans who began operating in Canada at this time were J.C. Hughson, of Albany, and Lewis Hotchkiss, of Ansonia, Connecticut.

Hughson bought timber limits in Haliburton, logs from which he ran down the Trent system to Lake Ontario, for export to his mill in Albany. Hotchkiss bought a sawmill in Collingwood. Hotchkiss and Hughson joined forces and bought the Muskoka mill in 1869. In 1871, Anson bought Hotchkiss's half-interest for about $40,000. The purchase covered an undivided half of the mills and all the timber limits. When Anson bought the half-interest, there were four mills — a lumber, a shingle, a lath, and a timber mill.[13] The Simcoe County *Directory* of 1871 reported a total production of 15 million board feet of square timber and sawn lumber, plus an unspecified quantity of shingles and laths. One hundred and twenty-five men were employed at the mills.

The first sawmill on Georgian Bay was built by the Upper Canadian government in 1830, at the mouth of the Severn River. Later called Port Severn, the village that grew up around it was known as Severn Mills. The mill, which was supposed to provide an income for the Indians living on the reserve near Coldwater, was built for the government by a contractor named Lewis at a cost of about £430 ($2,150). Regrettably, sawmilling had little appeal to Indians; they let the mill stand idle for several years and eventually put it up for sale. In 1837, the Indian agent at Coldwater reported that Chief John Asiance had been offered £75 for it by Indian trader Andrew Borland.

But Borland did not buy the mill. In 1836 Lieutenant-Governor Bond Head persuaded the Chippewa Indians to surrender the tract of land they occupied on the "public high road leading from Coldwater to the Narrows of Lake Simcoe."[14] By the terms of the surrender treaty, the Indians were to have been paid annual interest on the proceeds from future sales of the land to settlers. The Indians were subsequently moved to reserves — at Rama, on Lake Couchiching; on Snake Island, in Lake Simcoe; and on Beausoleil and

Christian islands, in Georgian Bay. In 1842 the Indian chiefs wrote to Governor Sir Charles Bagot complaining that they had not received any payment for their land and making it quite clear that the grist mill in Coldwater and the sawmill at Port Severn were not considered to be included in the land sale. The chiefs considered these mills "as Indian property."[15]

Sometime between the date on which the above letter was written and 1850, the Hon. William B. Robinson — member of Parliament for Simcoe County, commissioner of public works, brother of Chief Justice John Beverley Robinson, and chief negotiator of the land sale in 1836 — acquired the mill from the Indians. No record of the transaction survives, but Robinson also obtained a lease for the crown reserve consisting of broken lots 19 in the twelfth and thirteenth concessions of Tay township, which had been set aside for the Indian sawmill. James Sanson Jr of Orillia operated the mill for Robinson for a couple of years. Then, in 1852, Robinson withdrew, giving a seven-year lease of the mill site, "mill houses other buildings [and] improvements thereon erected"[16] and 150 acres to Sanson for £400.

Sanson's mill had only a single saw, with a cutting capacity of about one thousand board feet in a twelve hour period. Despite the mill's limited capacity, Sanson seems to have made substantial profit from it during the boom years 1853 and 1854: he was able to discharge the mortgage a full year before it was due. It was during the years that Sanson operated the mill that schooners began freighting lumber to American markets and the name of the village was changed to Port Severn.

The depression of 1857 induced Sanson to lease the mill site and property to Alexander R. Christie and Andrew Heron, merchants from St Catharines, Ontario. Christie and Heron obtained a licence for fifty square miles in Baxter township, commencing at Port Severn and extending east to the vicinity of Big Chute. Despite a depressed market for lumber, they made considerable improvements to the mill. The dams were extended and heightened, and the cutting capacity of the mill was substantially increased. A control dam for regulating the water level in the Severn River was built at Six Mile Lake. By 1862, the mill had two gang saws, one single saw, and two edgers. On the main island of Port Severn, then known as Sanson's Island, there were three houses, a store, and a stable; the mainland had a blacksmith's shop, more dwellings, stables, and a Catholic church.

Just as Kelly had overextended his resources in enlarging the Muskoka Mills, so Christie and Heron experienced financial difficulties at Port Severn. Consequently, they were either unwilling or unable to pay £517 ($2,585) owed to the contractor, Thomas McCormick of Barrie. McCormick obtained a writ of venditioni exponas from the Court of Queen's Bench, on the authority of which the sheriff of Simcoe County seized the mill and put it up for sale. The sheriff accepted an offer of only $1,000 for the Port Severn mill from Peter Christie, Alex Christie's son. A month after the sheriff's sale, Heron and Christie sold Peter the rest of the property not included in the sale, for £700 (about $3,400). Soon afterward, Alex Christie retired to Toronto, and the mill, then known as the Christie Mill, was operated by Peter.

On 7 April 1869 the Port Severn mill burned. Arson was suspected[17] but never proved. From the Etna Insurance Co. Christie collected $6,000 in insurance, with which he rebuilt the mill. The work was barely finished when Anson arrived and offered to buy Christie out. Typically, the price paid by Anson — $118,970 — was much in excess of the property's real value. He acquired the mill, wharfs, and equipment at Port Severn, fifty square miles of timber limits on the Severn River, and 9,966 acres of pine lands scattered throughout Simcoe County, in the townships of Matchedash, Medonte, Orillia, Tay, and Tiny . He also took over timber agreements on 3,462 acres that Christie had made with settlers. Anson paid $76,000 in cash and borrowed the remaining $42,970 from Alex Christie and William Kerr of Toronto, to whom he gave a mortgage at 7 per cent interest, payable by 1874 in four annual instalments. As with the other mills he puchased, Anson enlarged the Port Severn mill and village.

The picturesque village of Waubaushene, situated on a hill overlooking the entrance to Matchedash Bay, the southwest extremity of Georgian Bay, would eventually become headquarters and site of the principal mill of the Georgian Bay Lumber Co. Although located on a protected channel, the site was not ideal for extensive lumber operations in the mid-nineteenth century, when the only way of transporting lumber was by schooner or steamboat. There was no natural harbour, and the shallow water along the shoreline made the loading of barges and schooners difficult.

A network of trestles was built out to deep water to facilitate loading, but, even with these, some larger schooners could be only partially loaded. The half-filled vessels had to be towed to deeper water and loaded to full capacity from barges. After the Midland Railway reached Waubaushene, in 1875, the shipping problem was eased somewhat, although for several years afterward, water transport — the cheapest method   continued. In the latter years of the company's existence, when all the pine had been exhausted in Simcoe County and Muskoka and Parry Sound districts, sawlogs had to be towed at great risk and expense from as far away as Blind River on the north shore of Georgian Bay.

An Indian name, meaning probably "land of the rocky marsh," Waubaushene owes its existence to William Hall of Hamilton, who built the first sawmill there. There had been some settlement in the vicinity ever since 1834, when Catharine Fraser received a patent from the crown for lot no. 10 in the eleventh concession of Tay township. The 200-acre lot, divided first into halves and then into quarters, changed hands several times before 1861, when Hall bought the northern half of it from Preston Hallen, also from Hamilton, for $400.

Hall built a small sawmill in 1861. He apparently borrowed at least part of the capital from Edward Ferguson of Hamilton, because Ferguson held a mortgage on the property for £1,200 (about $5,000). In 1865, Hall bought the southern half of lot no.10 from Archibald C. Thompson of Barrie for $80 and all sixty acres of lot no. 11 for £30 from the Canada Co., which had acquired a crown patent to the lot in 1834. It seems that Hall enlarged the mill at this time; records in the Land Registry Office in Barrie show that Thomas Cundle and William D. Ardagh, both of Barrie, held mortgages of $6,000 and $20,000, respectively, on the property.

In July 1870, Anson paid Hall $39,000, in cash, for the Waubaushene mill, lots no.10 and 11, and timber rights to twenty-four square miles in Matchedash township. He paid Hall in addition $9,000 for 5,516 acres of pine land that Hall owned in fee simple, scattered throughout the townships of North Orillia, Flos, Medonte, Tay, and Tiny. Anson then spent a considerable amount enlarging the Waubaushene mill and village.

A thriving port — Tay Port — existed at the mouth of the Sturgeon River at the bottom of Sturgeon Bay long before the establishment of Waubaushene, two miles to the east. In 1844, the Sturgeon Bay Road, used for many years as a portage road between Georgian Bay and Orillia, was built between Coldwater, the former entrance to

The Port Severn Mill c1890   – courtesy C.A. Stocking

The Waubaushene Mill c1883   – Simcoe County Museum 369

the portage, and Tay Port. In the expectation that the harbour would grow to prominence, the government, in 1846, laid out a town plot — Port Powell — on lots no.9 and 10 in the ninth concession of Tay township. For a time, steamers and other vessels made regular calls, but the choice of Collingwood as terminus for the Northern Railway diverted traffic away from Sturgeon Bay, and Port Powell did not develop.

The first sawmill in the vicinity of Sturgeon Bay was built by Samuel P. Jarvis and Charles Thompson on the Sturgeon River, about one mile from the bay, on lot no.6 in the ninth concession. It is not entirely clear when the mill, known in the district as the Thompson Mill[18], was built, but probably about the time Port Powell was laid out. In 1836, Jarvis, chief superintendent of Indian affairs, bought lots no.6 and 7 from four veterans of the War of 1812, each of whom had received a half-lot — fifty acres — in a grant from the crown; Jarvis paid the veterans £30 each for their land. He and Thompson formed a co-partnership and built a water-powered mill which Thompson operated, but Jarvis held title to the property.

The mill seems not to have been profitable, for in 1851 lots no.6 and 7 were seized by the sheriff of Simcoe County and sold at public auction for unpaid taxes. The lots were purchased by Alexander McDonald of Toronto. Although McDonald and his successors would hold title to the land, Jarvis seems to have made an arrangement with McDonald about the mill, for in 1854 he gave a five-year lease of the mill to Robert Buchanan of Coldwater at an annual rental of £200.

The following year, Buchanan — son-in-law of pioneer contractor Jacob Gill and future father-in-law of W.J. Sheppard, later president of the Georgian Bay Lumber Co. — leased the mill to William Borland and Patrick Connor. Borland and Connor contracted to pay the annual rent and to deliver all the lumber produced to vessels at Port Powell, on behalf of Robert Buchanan, who agreed to pay them a fixed rate of 25 shillings per one thousand board feet. In 1856, Borland and Connor, for reasons unknown, assigned the agreement and lease to James Sanson, owner of the Port Severn mill. Sanson borrowed 600 pounds from his father, James Sanson Sr, with the intention of enlarging the mill. He mortgaged the lease and the agreement with Borland and Connor to his father as security on the loan.

This bizarre series of assignments came to an end with the depression of 1857. Sanson sold the Port Severn mill to Christie and Heron, but they did not take over the Sturgeon River mortgage, and there is no record of the mill operating after that date. In 1865, Alexander McDonald died. Lots no.6 and 7 were willed to his son John McDonald, who sold them to William Hall in 1867. The lots next came into the possession of Anson Dodge, in 1871, as part of the Waubaushene mill and land purchase from Hall.

Meanwhile, Jarvis had bought lot no. 8 in the clergy reserve sales of 1848. The lot, which spanned the Sturgeon River at its mouth, had a mill site on it. Sometime between 1853 and 1857, Jarvis erected a small, water-powered sawmill on the lot. Jarvis died in 1857, having willed the mill to his son, Charles Frederick.

In 1867, the executors of Jarvis's estate, with the concurrence of Charles, sold lot no. 8 with the mill to William Laramy and Alva Smith of Batavia, New York, who operated under the firm name of Laramy and Co. Laramy and Smith also bought the west half of lot no. 9, on which they erected a steam sawmill, mortgaged to the Bank of Toronto for $20,000. The partners apparently intended to lumber on a large scale, for, in 1869, they obtained timber licences for 108 square miles on the Severn River in the townships of Baxter, Morrison, and Wood. In 1871, Smith died. Laramy had difficulty meeting the mortgage payments, and the bank seized the mills and timber limits.

Anson decided to add the Sturgeon Bay mills to his Waubaushene and Port Severn operation, and so, in the spring of 1871, he bought the two mills and the Severn River timber berths from the bank and Laramy for $42,000. This purchase was probably the best of all the ones he made, not because of the mills, with their limited cutting capacity, but for the large area of valuable timber that was included in the purchase.

It became inevitable that the port of Collingwood would develop into a major grain depot and lumbering centre when the spot, situated opposite a group of rocky islands known as the "Hen and Chickens," was chosen as the terminus of the Northern Railway on Georgian Bay. Before the selection of the terminus was made public, Benjamin W. Smith, sheriff of Simcoe County, learned about it through official county sources. On the strength of this confidential information, Smith acquired crown patents for lots no.43 and 44 in the ninth concession of Nottawasaga township and had them surveyed into building lots. Operating covertly, Smith joined David Morrow and the Reverend Mr Lewis Warner as silent partners of Joel Underwood, a blacksmith who had come to Nottawasaga township from the United States in 1847.

The partners immediately began to develop a townsite, in the spot where they knew the railway would reach the shore. In 1852, Underwood built a steam sawmill at the mouth of Underwood Creek and opened a store on First Street. The mill, the store, and the dwellings that were built near the mill became the nucleus of the future town of Collingwood. Because the mill was built two years before the railway reached Collingwood, the boiler and machinery had to be hauled in by team.

In 1867, Benjamin Smith gave a ten year lease of the mill site, with an option to purchase, to Lewis Hotchkiss and Stephen J. Peckham. In 1869 they bought the mill for $7,000 and rebuilt it.

Peckham, his son Jesse, and his sons-in-law Edwin Stocking and Isaac Hoag — Quakers from New York state — operated the mill in co-partnership with Hotchkiss under the firm name of Hotchkiss, Peckham and Co. The partners owned several thousand acres of timber land, in fee simple, on the shores of Nottawasaga Bay, in Flos, Nottawasaga, Sunnidale, and Tiny townships. It was reported in 1870 that the company employed 140 men in its lumber, shingle, lath, and planing mills. The 150-horse-power steam plant was capable of producing daily 150,000 board feet of lumber, 40,000, laths and 12,000 shingles. But unless timber berths were obtained elsewhere, the company had a limited future, because the standing pine on its Simcoe County properties was rapidly diminishing. The partners decided to sell.

In August 1871, Anson Dodge bought the mill and timber lands from the partners for $170,000, assuming existing mortgages on the property, amounting to $64,000. Anson agreed to pay Peckham, Hoag, and Stocking $85,000, in annual instalments, at 7 per cent interest. Hotchkiss was paid $125,000, which included $40,000 for Hotchkiss's half-interest in the Muskoka Mills property which Anson purchased at the same time. Also, Anson bought the steam tug *George Watson* from Hotchkiss and Peckham for an additional $3,000 in cash.

It was Anson's intention to supply the Collingwood mills from timber berths he had acquired in Muskoka and Parry Sound, after the pine around Nottawasaga Bay was exhausted. With the purchase of the Hotchkiss and Peckham mill, Anson then owned out-

The Collingwood Mill c1885 — courtesy Robert Thiffault

The Longford Mill — Ontario Archives 3177 #54

right, or held a major share in, all the sawmills on Georgian Bay from Collingwood to Byng Inlet, except one: the mill owned by Kean, Fowlie and Co. at Victoria Harbour. Only time would tell if his investment "insanity" would bring about his ruination, or make him "one of the wealthiest men on the continent."

Not satisfied with the eight mills located on Georgian Bay, which tapped the vast pine resources in Parry Sound and Muskoka districts and Simcoe County, Anson also sought access to the pine in Haliburton district and Victoria and Ontario counties. Thus, in 1870, in partnership with John Thompson, he built a steam sawmill near Longford in Rama township, on the strip of land that separates Lake Saint John and Lake Couchiching. The unincorporated partnership, operating under the name of Thompson and Co., had timber holdings in Anson township in Haliburton; in Carden, Dalton, Digby, and Longford townships in Victoria County; in Rama township in Simcoe County; and in Thorah township in Ontario County. Most of the sawlogs and square timber were floated down Black River and Log Creek into Lake Saint John, on which the mill was located, on an eighty-acre site on the west side of the lake. Logs for the mill were carried from the lake under a bridge on the Rama Road on an ingenious steam-operated conveyor. Square timber and sawlogs for export were carried a further 750 feet on the same conveyor to Quarry Bay on Lake Couchiching, where the firm had constructed docks and piling facilities.

The mill was ideally situated, being near stands of pine in Haliburton and Victoria counties. Lumber was easily towed in barges by steamer, and square timber and sawlogs in booms, to the Northern Railway depot at Belle Ewart or the Midland Railway terminus at Beaverton. From Belle Ewart, lumber, logs, and timber were shipped to Toronto, for export by water to Albany and Cleveland; from Beaverton, the wood was shipped to Port Hope. Also, when the mill was built, Anson knew that the proposed Muskoka Junction Railway would pass right through the Longford mill property.

Chapter Three

Timber and Transport

Anson wanted to secure the large investment he had made in saw-mills and property on Georgian Bay by ensuring that there would be enough pine behind his mills to last for at least eighteen years. Most of the large pine suitable for squaring had already been cleared from the property in Simcoe County acquired by Anson with the purchases of the Collingwood, Port Severn, Sturgeon Bay, and Waubaushene mills. Smaller trees, suitable for sawing into lumber, would last only a few more years, given the rate at which his enlarged mills were sawing them.

The pine needed to supply his mills would have to come from virgin forests in Muskoka and Parry Sound districts. In fact, it was his knowledge that pine would be available for licensing in these districts that brought Anson to Canada in the first place and encour-aged him to invest heavily in sawmills on the shore of Georgian Bay. Therefore, concurrent with acquisition of the mills, he set about obtaining as many timber licences as possible.

One of the first major policy decisions taken by the post-Con-federation Ontario government was to increase the population of the young province. To this end, the Free Grant and Homesteads Act was passed in 1868, primarily to attract British settlers to Ontario but also to counteract the US Homestead Act, which was also encouraging Britons — and even Ontarians — to go to the United States. It was hoped, moreover, that the Ontario act would speed settlement of the sparsely populated northern part of the province with hardy pioneers, capable of developing the north's rich resources, needed by Ontario to lay the foundation for a pros-perous future. A large tract of land lying between Georgian Bay and the upper Ottawa River — the Ottawa-Huron Tract — was set aside for settlement. The first townships selected for free grants were located in Haliburton, Muskoka, and Parry Sound districts.

To build up revenue in anticipation of the enormous expenditure of public funds for roads and public buildings in the new settlements, the Ontario government substantially increased timber dues (the principal source of provincial revenue) and put new timber regulations into effect. To ensure collection of vital forest revenue, the commissioner of crown lands employed travelling inspectors, or "woods rangers," to visit lumber camps to verify the measurement of timber cut and determine the amount of dues payable.

Lumbermen paid for the privilege of cutting timber in three ways: by a charge for the timber licence, by annual ground rent, and by dues on the timber and lumber actually cut. After 1866, timber licences were sold at public auction with an upset price, or reserve bid, determined by the commissioner of crown lands. Prior to a sale, timber berths were laid out and surveyed by government timber cruisers, who determined the quantity of timber in each berth. Based on the current level of dues, an upset price for each berth was established. Any sum obtained for a berth at public auction above the upset price was known as a bonus. When the market was strong, as it was in 1871, valuable timber berths brought considerable bonuses. Two dollars per square mile was charged, annually, for ground rent, and timber dues were fixed at $25 per thousand cubic feet on squared timber and $1.50 per thousand board feet on saw logs.

Before opening the north for settlement, the government considered it advisable to offer for sale, by public auction, timber limits in the grant lands in selected northern townships. The first public auction of timber limits in Muskoka and Parry Sound was held at the Parliament Buildings in Toronto on 23 November 1871. Anson was one of seven lumbermen to attend.

He was well prepared. Prior to the sale, he hired his own timber cruiser to estimate the value of the timber in the berths on which he intended to bid. Typically, the man Anson hired was one of the most knowledgeable timber cruisers in the business. He was Aubrey White[1], a young Irishman who had come to Canada in 1862 at the age of seventeen and had worked in the lumber industry for several years.

Just before the sale, Anson learned that the Cook brothers, who were building a mill at Victoria Harbour, were planning to bid on the same limits. Anson and the Cooks decided to share the territory. Rather than bid against each other, each agreed to bid on half of the berths and to work out later their own arrangement for sharing the timber. Between them they bought twenty-seven timber berths, encompassing 359 square miles, for which they paid $87,830 — an average of $245 per square mile — in bonuses. Added to the timber lands and licences he acquired with the purchases of his mills, the new timber limits would supply pine for Anson's mills for many years to come. He had fallen far short of owning "⅛ of the standing timber in Canada," but he now owned a sizeable piece of it and possessed the wherewithal to become rich.

Anson participated in planning and promoting two railway companies in Ontario — the Toronto, Simcoe and Muskoka Junction Railway and the Lake Simcoe Junction Railway. When he arrived in Ontario, businessmen in Toronto were already talking about the desirability of building a line from Barrie to Orillia to connect with the Northern. Anson joined the group. He soon convinced the others that the proposed route should be extended beyond Orillia into Muskoka, to provide access to the sawmills and timber limits he had already bought or was planning to buy.

In September 1869, Anson conducted a large and influential deputation of officials of the Northern Railway and other prominent men through the districts of Muskoka and Parry Sound, so that they could see the advantages of investing in a railway into the north. In the evening of 6 September, the party left Gravenhurst on the steamer *Wenonah* and travelled to Bracebridge, the developing business centre of Muskoka district. Next day they returned to Gravenhurst, where a public dinner was held in the evening. While no firm promises were given, remarks made by officials at the dinner hinted that a railway might be built if the districts provided some of the construction capital.

On the ninth, the party left for Parry Sound on *Wenonah*, following the route Anson had taken in 1866 and had followed several times since. After a pleasant picnic lunch on an island off Port Carling, they cruised up Lake Rosseau on the new steamer *Wabamic* to the village of Rosseau, where they spent the night in Mrs Irwin's hotel. Next morning they travelled in four stage-coaches to Parry Sound, arriving about four o'clock in the afternoon. Most of the villagers turned out to greet them. To mark the occasion, the village was festooned with flags and banners, while the steamers and sailing vessels in port vied with each other as to which would make the grandest show. Anson's tug *Mittie Grew*[2] flew the "Stars and Stripes" beside the Union Jack. Next morning the party was conducted on a tour of the village to inspect the mill, the harbour, and other points of interest. In the afternoon, it travelled by wagon along the Northern Road to view some of the falls on the Seguin River. That night a gala banquet was held in the Seguin House Hotel, at which speeches were delivered by Anson, the Hon. John Beverley Robinson, Mayor Harman of Toronto, and Fred Cumberland, the managing director of the Northern Railway. The speakers extolled the romantic scenery they had seen and commented on the beauty of the lakes and rivers, but, although railways were talked about, no commitment was made to build one to Parry Sound.

The Muskoka — Parry Sound tour seems to have had the desired effect, because the promoters decided that the projected railway from Barrie to Orillia should be extended into Muskoka; hence the railway's name. The first organizational meeting of the provisional board of directors was held in the office of the mayor of Toronto on 29 October. Anson Dodge was elected vice-president. On 24 December, the Toronto, Simcoe and Muskoka Junction Railway came into being, when royal assent was given to the provincial statute[3], which authorized the company to build a railway not only from Barrie to Orillia but also to an unspecified terminus on Lake Muskoka.

The act approved construction of a railway through the counties of Simcoe, Ontario, and Victoria. To accomplish this, the road would have to cross the narrows at Atherley and run up the east side of Lake Couchiching, right past the Dodge and Thompson sawmill at Longford. We do not know whether Anson influenced the directors' choice of the longer route around Lake Couchiching to Washago, because he had already planned the Longford mill, or whether he built the mill at Longford, knowing the railway would pass there; but it seems clear that the two events were connected.

For the next two years, the railway project lay dormant, sufficient capital for construction not being available. In February 1871, the merchants in Bracebridge, on learning that Anson would pass through the village on his way to Byng Inlet, planned a breakfast at the Dominion House Hotel in his honour. Ostensibly given to provide an opportunity for the citizens of Bracebridge to testify to the esteem in which they held Anson, the breakfast was actually intended to remind him about the railway and to request that he "use [his] influence to hasten the day when [the citizens] would be blessed with railway facilities."[4] Anson promised to do what he could.

He seems to have kept his word, for on 1 April 1872 the railway was opened for traffic between Barrie and Orillia. From Orillia, the track crossed, as planned, to the eastern shore of Lake Couchiching, and it was completed to Washago, at the head of the lake, on 8 August 1873. Construction slowed down considerably because of the depression, and so the line did not reach Severn Bridge until 1874 and, finally, reached Muskoka Wharf in November 1875.

Another line, the Lake Simcoe Junction Railway, which was to run from the village of Sutton on Lake Simcoe to Stouffville on the Toronto and Nipissing Railway, would not contribute significantly to Anson's lumbering enterprise. His interest in investing in this railway seems to have been to share in the profits the line was expected to generate. The act[5] incorporating the company received royal assent on 29 March 1873. A first meeting of the provisional directors, of which Anson was one, was held in the company's office, on Church Street, Toronto, on 7 May. Construction was stalled for a time because of the worldwide depression; consequently, it was not until October 1877 that the road was opened to Jackson's Point, a short distance beyond Sutton, on Lake Simcoe. As predicted, it was a highly profitable railway.

Anson supervised his extensive lumber empire from the magnificent estate he established at Roche's Point (then called Keswick) on the south shore of Lake Simcoe. The land and house around which he built were originally owned by the Reverend Mr Walter Stennett, a former principal of Upper Canada College, in Toronto. When Stennett retired from the college in 1862, he obtained a crown grant for a piece of land in North Gwillimbury township at the mouth of Cook's Bay, directly opposite Belle Ewart. In 1869, Anson bought the property from Stennett for $8,000. He added other pieces of property until the estate comprised about 250 acres.

Anson spent a good deal of money in the next three years enlarging and renovating the house, building a dock and boathouse, and landscaping. "No mortal could possibly know how much I spent,"[6] he later complained with some bitterness; on reflection, he estimated that over $50,000 had been spent — much of it during his election campaign in 1872. The three-storey house, which Anson called Beechcroft because of the magnificent beech trees that once surrounded it, is still in use. It has twenty-five rooms, including fourteen bedrooms. Constructed of field-stone and covered with stucco, the walls are one-and-a-half feet thick. The house has a southern look about it, undoubtedly designed to appeal to Anson's Virginia-born wife. With its gabled roofs and dormers, bay windows and encircling veranda, supported by y-posts, Beechcroft would not look out of place in the Virginian countryside. An iron weather-vane standing on the roof of the boat-house has a "D" incorporated into it, recalling the name of the man who had it placed there.

Anson employed Frederick Olmstead, the designer of Central Park in New York, to lay out the grounds. Black earth was barged in from Holland Marsh to enrich the sandy ground around Roche's Point. A steam-heated "grapery" and a hot-house were built, to ensure an early start for bedding-out plants. Subsequent subdivisions of the original estate have destroyed much of the overall plan of the grounds, but beautiful Norway spruce planted in conjunction with the wide driveways survive and are a prominent feature of Beechcroft. As protection for the young spruce trees, fast-growing Lombardy poplars were planted, and they remained a distinguishing characteristic of the estate's boundaries until the 1920s. Large indigenous elm trees, once plentiful on the estate, have also disappeared, victims of Dutch elm disease. A fifty-eight-acre lot adjoining the rear of the original estate was purchased by Anson for $5,000 and converted into a park. An eight-foot-high log fence was built around the lot to enclose a small herd of deer. Today most of the original park is a pine forest; only the street name — Deer Park Drive — recalls its one-time existence.

Beechcroft was ideally located for Anson's purposes. While providing the advantages of country living, it was situated near enough to Toronto for easy access. In Anson's day, one travelled the forty miles to Toronto by taking a boat to Belle Ewart and the train into the city; or one could travel to Newmarket by stage-coach and take the train from there. Anson had a steamboat in which he travelled up and down lakes Simcoe and Couchiching to and from his various mill locations. From Belle Ewart, he could take the train to Collingwood, and thence to any point on Georgian Bay by steamer. He could travel in his steamboat directly to his companies' temporary headquarters in Barrie or to his sales office in Orillia; Waubaushene, Sturgeon Bay, and Port Severn were only a few hours by stage from Orillia. Longford was accessible by boat, and he could travel on to Washago and into Muskoka and Parry Sound districts by the routes described earlier. A very active man, Anson visited his mills frequently, especially when he was purchasing property and expanding the mills' operations.

Beechcroft Roche's Point

Letterhead – Georgian Bay Lumber Company of Ontario - 1873

Chapter Four

A Deo Victoria

On 13 February 1871, letters patent were issued, under the great seal of the Dominion of Canada, incorporating the Georgian Bay Lumber Co., with capital stock of $229,500. The company, which included only the Port Severn and Waubaushene mills, was empowered to carry on the business of lumbering in all its branches and to establish "a line of Steamboats or Sailing Ships or vessels from and between Port Severn and the cities of Chicago and Cleveland . . . for the purpose of carrying lumber and other freight between said ports."[1]

The headquarters of the company was to be in Orillia; other centres of operation were Waubaushene and Cleveland. Anson acquired 2,000 shares (worth $100 each); W.J. Macaulay, manager of the Orillia office, took 250 shares; John Beverley Robinson was given five shares; and the other forty-five were divided among Alexander Christie, William Kerr, D'Alton McCarthy, and Daniel Sprague. Anson was president, Sprague was bookkeeper and treasurer. The provisional directors were Anson, Macaulay, and Robinson.

After he bought the Sturgeon Bay property, Anson decided to reincorporate the Georgian Bay Lumber Co. Because the province, not the Dominion, issued the vital timber licences, he thought it prudent to apply for a provincial charter. Thus, on 2 March 1872, he was granted a charter, by provincial statute, for the Georgian Bay Lumber Co. of Ontario[2], with capital stock of $1 million, in shares of $100 each. Anson held 9,880 of the 10,000 shares issued. The remainder (120) were divided among the same men who owned shares in the company incorporated by Dominion charter a year earlier.

The provisional directors and other shareholders held a series of meetings in Orillia, in April, to organize the company. Anson was elected president, and W.J. Macaulay became vice-president. Anson, Christie, and Macaulay were elected directors. Macaulay and D.W. Linn (formerly manager of Sage's mill at Belle Ewart) were appointed general managers, Linn having the added responsibility of treasurer. D.E. Sprague was made assistant manager, with temporary headquarters in Orillia. Levi Miller became manager of logs and woods operations, and R.W. Brewster was placed in charge of sales. On 3 May 1872, the Ontario firm bought the Dominion-chartered Georgian Bay Lumber Co. for $160,374.97, but this was probably a paper transaction only.

The company was overcapitalized. At the time of incorporation, it owned only the Port Severn, Sturgeon Bay, and Waubaushene mills and the timber limits held by the previous owners of these mills, all of which Anson assigned to the new company for $185,000. Because he intended the Georgian Bay Lumber Co. to be the hub of his extensive lumber operations in Canada, Anson planned to add to the company's holdings the Collingwood, Longford, and Muskoka mills and the large tract of timber that he purchased at public auction in 1871. But even with these properties and the amount spent on enlarging the mills, Anson had invested less than $600,000 in the properties that were to have been controlled eventually by the Georgian Bay Lumber Co. of Ontario. He believed the $1 million capitalization justified, however, because he predicted an increase in the value of his mills and timber limits and expected to have a large inventory of lumber and timber always on hand. Like most of Anson's predictions, this one proved overly optimistic.

The head office of the Georgian Bay Lumber Co., and indeed the head offices of all of Anson's lumber companies, were initially in Barrie. His small headquarters staff operated out of a suite of rooms in the McCarthy Block, leased by D'Alton McCarthy to Anson for $140 per year. It made sense to locate the headquarters in Barrie: McCarthy's practice was there, and he was solicitor and secretary of each of Anson's companies; moreover, Barrie was accessible to Toronto by rail and Beechcroft was only a few miles away by water, a factor that may have led Anson to build his home at Roche's Point. If Barrie was the administrative centre of the companies, the mill village of Waubaushene was the operational capital of his vast lumbering empire; in 1873, the head offices of all the companies were actually moved there.

Little is known about the mill village of Waubaushene during the years that Hall owned the mill. Presumably the Halls — Elizabeth and William — had a house at Waubaushene, and there would have been a boarding-house and dormitory for the mill hands. The village had a Roman Catholic church, built by Jesuit Father Theophile Francis Laboureau, who came by water from Penetang to conduct masses at both Waubaushene and Port Severn. This little church — twenty feet by thirty feet — was erected on a small lot — sixty feet by fifty-three feet — donated by Hall to the Roman Catholic Episcopal Corporation of the Diocese of Toronto. It is not known when the church was built, but it was there in the spring of 1870, when Anson bought the mill. The church property and public access to the water's edge were excluded from the sale. Hall subsequently deeded the lot to the diocese for one dollar.[3]

Sufficient information is available about Anson's brief proprietorship of Waubaushene to provide some insight into the corporate approach that he established. This view of things when put into practice by his senior empoyees and their successors, would determine the modus operandi of the Georgian Bay Lumber Co. and set the social patterns and conventions of its logging camps and mill villages, particularly Waubaushene, for the next fifty years. Briefly stated, it was Protestant, paternalistic, temperate, and ascetic; it was also marked by fairness, equity, and a spirit of co-operation.

But Anson did not originate the approach under which he operated, any more than a fish could create the water in which it swims. He was born to it. His management "philosophy" was a product of the American social character and work ethic of the age and more especially of the family that produced him. The Phelps Dodge business views to which Anson subscribed, at least superficially, had their roots in the Protestant ethic of an earlier period.

The Protestant ethos of late-seventeenth-and-early eighteenth-century America was an expression of Calvinist and Quaker individualism and asceticism. It was a religious imperative, a calling to work for the glory of God. It reinforced traits of rigid self-discipline, saving, and deferred rewards. It demanded of its followers constant work at a vocation or "calling" as proof of faith.

The Puritan was compelled to prosper, to become rich for God, though not for the enjoyment of the flesh. This ethic supported the development of a social character that produced the ascetic, rugged individualist who was disciplined, productive, and self-righteous and who sought to overcome all doubts about salvation through prayer and hard work. This was the work ethic that motivated many successful crafts people and allowed them to adapt readily to the kind of self-sufficient business that could be run by one person, such as a farm or a workshop.

The shutting off of imported English goods during the Napoleonic Wars, followed by an even greater restriction of trade with Britain during the War of 1812, started the United States on the road to economic self-sufficiency and stimulated the rise, eventually, of a vast industrial economy. The switch from commercial, agricultural, and craft industries as the chief sources of livelihood to an industrial-based economy would eventually produce a new work ethic and a change in the social character of the United States.

As the economic situation changed with the advent of the small businessperson, the entrepreneurial ethic emerged. The entrepreneur exchanged the craftsperson's traits of caution and moderation for daring and speculation. Because the entrepreneur-industrialist could achieve only through the efforts of others, the new social character required that the strong, egotistic individualism of the crafts give way to a spirit of social co-operation.

How to obtain the co-operation of workers so as to ensure a high level of productivity is a problem that has preoccupied managers and intrigued industrial psychologists since large organizations were invented. Early-nineteenth-century entrepreneurs, like Phelps and Dodge, sought to temper the individualism of the former age and produce the co-operative and conforming workers required by the new industrial society by appealing to religious bonds of community. Evidence of this can be seen in the emphasis placed on religious education in the school curriculums of the day and the growth of the Sunday School movement. Although the frontier would later inject a more ambitious and reckless entrepreneurial spirit into the American social character, the values of the independent farmer-crafts-person-small businessperson remained the dominant ideal until after the Civil War. It was the craftsperson's work ethic, overlain with an emerging nineteenth-century entrepreneurial drive, that marked the Phelps Dodge business viewpoint and ultimately that of the Georgian Bay Lumber Co.

The life of Anson Greene Phelps, Anson's grandfather, spanned the two periods described above, and his career reflects this. He was born in 1781, the year the American Revolution ended. Orphaned at a young age, he was raised by a minister friend of the family. In 1799 he entered an apprenticeship under his brother in Hartford, Connecticut, to learn the saddler's trade. At the end of his training, he established his own shop in Hartford, where he made and sold saddles, trunks, and other articles of leather. But young crafstman Phelps was a business expansionist with the shrewd ability to see and exploit business opportunities; soon his shop's merchandise was expanded to include groceries, crockery, and flour.

He next found a market for his saddles and harnesses among the plantation owners of South Carolina. Doing business in the south introduced him to the cotton trade, and vey soon he was shipping cotton to England and importing the products of English mills and consignments of iron, copper, tin, and other metals in demand in the United States. As business expanded, Phelps bought first a few small sloops, then a number of square-rigged coastal packets, and finally he became a shipowner and shipping agent on a large scale. In a few years, Phelps, the craftsman, had become Phelps, the entrepreneur.

In 1812, Phelps moved his operations from Hartford to New York. In 1824 he formed a partnership with another New England entrepreneur, Elisha Peck. For the next ten years, Phelps and Peck carried on a successful, diversified export and import business with England, chiefly in cotton and metals.

Meanwhile, Phelps's two oldest daughters, Elizabeth and Melissa, had married two young merchants, Daniel James and William Earl Dodge. Because his sons-in-law had more or less the same ideals, convictions, and religious points of view as he, Phelps proposed that the three of them form a family partnership and take over the business of Phelps and Peck. The young men agreed, and in 1834 Phelps, Dodge and Co. was formed.(In England the firm operated under the name of Phelps, James and Co.) Phelps Dodge soon expanded into lumbering, into railways and finally into mining and metalurgy, eventually becoming one of the largest copper-producing and copper-fabricating companies in the United States.

Messrs Phelps, Dodge, and James were all bred in the Puritan tradition. As Presbyterians, they accepted the stern, uncompromising creed of John Calvin and John Knox. Consequently they believed that one must be both diligent in business and fervent in spirit. In his *History of Phelps Dodge*, Robert Cleland summarized the oneness of the partners' religious convictions, their public responsibilities, and their business practices:

> Men like Anson Phelps, and there were many of his Puritan upbringing in that day, looked upon the maintenance of the church and its agencies as both a privilege and an inescapable obligation. The support of properly directed philanthropies and educational institutions was also as obligatory as the care of one's own family. Frugality, charity, respect for private property, obedience to law, responsibility for the exercise of a man's talents in business, as well as in every other phase of life — these were the by-products, good or bad, of that rugged, sometimes intolerant Puritan-Presbyterian tradition, a tradition whose, strong tough threads were woven into the very warp and woof of the partnership that in the mid-thirties took over the business of Phelps and Peck and magnified it into one of the greatest business organizations of its time.[4]

The business philosophy of Phelps, Dodge, and James was not merely a declaration of purpose to appear in an annual financial statement or a summary of goals to be tacked on an office wall. For these men, business and personal goals were one and the same; they believed in them ardently and pursued them religiously. Nowhere was their corporate philosophy more rigorously applied, and nowhere were the "by-products, good or bad, of [their] rugged, sometimes intolerant Puritan-Presbyterian tradition" more evident than in their mill villages and factory towns. The by-product was good: the villages were safe, clean, and well managed; the living accommodations were "modern" and generally free; all the physical, medical, educational, and religious needs of the workers were provided for. If the by-product was bad it was so only because the philosophy was highly paternalistic. A condition of employment in a Phelps Dodge mill was total submission of will to the company — no alcohol, no union, no strikes, no voice in the organization or management of the community. Regular attendance at church was expected.

It was William Earl Dodge who directed the energies of the company into lumbering. Just a year after the partnership was formed, Dodge, who "seemed to comprehend the vastness of the lumber interests in Pennsylvania,"[5] began purchasing timber lands there. The first land purchased was in Tioga County, and the first mill was erected in the village of Wellsborough, later renamed Ansonia. In November 1836, after an inspection of Wellsborough, Anson Phelps noted in his diary that there was "no church nor stated preaching of the gospel"[6] in the village. "I hope the Lord will enable me to do something towards supplying the want,"[7] Phelps pledged to his diary. The Lord co-operated, and four years later the diary reveals that Phelps and a companion journeyed to Wellsborough to dedicate a Presbyterian church. William Earl Dodge donated a large Bible, and in 1844 Mrs Dodge purchased a church bell in France and had it shipped to Wellsborough. (The church, the bell, and the Bible are still there.) Religious and temperance tracts were regularly sent from the company's head office on Cliff Street, New York, and distributed among the employees.

The partners built a flour mill, erected a "company store," provided a school, and operated boarding-houses for the workers. The pattern set at Wellsborough was followed in other sawmill villages and in the copper and brass manufacturing town of Ansonia, Connecticut. Later it was repeated in the mining towns of Arizona.

One of the showcase mill villages of Phelps, Dodge and Co. was on St Simon's Island in Georgia. The mill and outbuildings, church, school, store, and houses for the employees formed an attractive village that was admired by travellers taking the inland steamer route to Florida. Another model mill village was Waubaushene, established in the Phelps, Dodge tradition by Anson in 1871.

When Anson signed on to apprentice in the lumber trade with Henry James and Co. in Baltimore, he learned more than just how to grade and market lumber and how to keep accounts; he was thoroughly schooled in the Phelps-Dodge-James business viewpoint, the principal tenets of which he had already learned in his father's house. But Anson was not willing to accept completely the paternalistic, austere ethos that motivated his father and uncles and, before them, his grandfather. Since boyhood he had rebelled against the domineering creed of his father; although he would imitate it, he had no desire to apply such a creed, literally, in his own business practice. His journeys abroad and his restless wandering about the United States had brought him into contact with the ideas of great liberal thinkers of the past, from Montesquieu and Rousseau to

Thomas Paine and James Madison. He tried to incorporate something of their thinking into his own business thought and practice.

The seal of his Georgian Bay Lumber Co., 1871, which he crafted with great care, and which was described by Thomas McMurray, publisher of the Bracebridge *Northern Advocate,* as "appropriate in design and chaste in finish,"[8] reveals a good deal about how Anson intended to manage his mills and mill villages in Canada.

In the middle of the seal is a shield, and in the centre of the shield is a pine tree, representing not only the product of the company but symbolizing, also, the efforts of the managers and workers. Entwined around the tree are the initials A.D.V., standing for "A Deo victoria," the principal half of the company's motto, in which the tree is rooted. This motto — victory to God — reminds us of the Puritan-Presbyterian ethic in which Anson's own roots were imbedded and which, like the companies of his ancestors, presumably underscored the Christian foundation of his own firm. But the tree reaches skyward, terminating in the other half of the motto — co-operation — without which the trees could not be harvested and God's ultimate victory, i.e. man's salvation, achieved. The shield is encircled with the name of the company and the date of incorporation.

It was in the way he hoped to acquire his employees' co-operation that Anson's approach to business differed substantially from his father's. Charity and paternalism were not enough. There was a kind of Rousseauean social contract morality in his approach to running a lumber business. Anson explained to the publisher of the *Northern Advocate*: "It is my intention to set apart a certain amount of annual profits derived from my business and create a fund which will be distributed among the principal clerks, foremen and head sawyers, thus making deserving men in my employ shareholders and gainers in my property. And so," he reasoned, "I will secure over my different departments the best men to be found, and in the fact of their having an interest in the concern will encourage them to do all in their power for mutual benefit."[9]

Profit sharing — and that is what Anson had in mind — originated in Paris, France, in 1842, where it was initiated by a wealthy house decorator named Leclaire. Anson had tried profit sharing with a degree of success in 1869, when he gave one or two shares of the second Dodge and Co. to his loyal employees Samuel Schofield and William Jay Hunt. He was one of the first to contemplate instituting the system in Canada. The publisher of the *Northern Advocate* was greatly impressed with Anson's plan: "From what we know of Mr. Dodge, we have every confidence that he will mature such plans for the good of those under his care, and will eventually draw around him[self] a noble class of men whose highest ambition will be to promote his interest; and the beauty of the arrangement is such that in doing so they will advance their own."[10]

For the workers, Anson provided the same kinds of facilities and amenities as existed in the mill villages of Phelps Dodge and Co. in the United States, but his reason for providing them was different. He was concerned not so much with saving workers' souls as with maximizing their efforts. Thomas McMurray of the *Northern Advocate* gives us some insight into Anson's motive:

> We are pleased to learn that Messrs. Dodge and Company, with a view to promoting the comfort and improving the condition of their men, have commenced to make great improvements at their several establishments. Churches, school houses and dwelling houses of a substantial and superior class are being erected at the Severn, Waubaushene, Byng Inlet and the Magnetewan, that the social, moral and intellectual welfare of the people may be provided for. We have always held the opinion that it was the duty of the employer and at the same time his advantage to take an interest in those under his care, and we have always noticed that those employers who exercise the greatest oversight towards their men prospered most — it invariably provides mutual benefit. Men, when cared for, take a deeper interest in everything that pertains to the welfare of their master, and if employers would study their own best interest they would carry out this principle. Messrs. Dodge and Company have certainly taken a step in the right direction and we trust that the men will appreciate the kindness shown by a faithful discharge of their duties.[11]

The business genius of the partners in Phelps Dodge and their successor sons was their willingness to employ competent specialists and then grant them the authority and flexibility to operate. The Dodges, for the most part (Anson excepted), were financial and investment wizards — corporate strategists. Other than for Anson Phelps's one-time expertise on a saddler's bench, and William Earl Dodge's earlier aptitude for stocking store shelves and dealing with customers, none of the partners had any particular technical skill;

they had no practical experience in the technological side of the several industies they would acquire. No Dodge had ever chopped down a pine tree, ridden a log down a treacherous river, operated a steam-engine, worked the levers that activated a saw, squared a timber, or piled lumber; but they hired others skilled in these operations and treated them well, and, as a consequence, the Dodges prospered.

Anson did not have an opportunity to implement fully his profit-sharing scheme. In none of the three or four years during which he operated his mills in Canada did he make a profit. Consequently, he was not able to establish a fund for distribution among "principal clerks, foremen and head sawyers," as planned.

He did, however, distribute shares in his several companies among senior officers — Christie, Dill, Hunt, Macaulay, McCarthy, Schofield and Sprague — and he gave shares to his political benefactors, especially Angus Morrison and John Beverley Robinson. But because there were no profits to distribute, these shareholders did not benefit from Anson's generosity. Ordinary workers benefited from the advantages derived from the "great improvements" at the several establishments carried out under the direction of the mill managers and technical staffs Anson brought from Pennsylvania.

The principal supervisors and technicians he brought to Canada from Pennsylvania to manage the day to day operation of the Georgian Bay Lumber Co. were James H. Buck, Theodore W. Buck, D.J.Cooper, John C. Else, Andrew Meneilley, Henry Milne, John R. Pierson , William H. Russell, and Philip Schissler. Several other technical people — sawyers, filers, steam engineers, storekeepers, and office managers — came up from the Pennsylvania operation either in 1871 or later, but there is no record of exactly who they all were or when they came. One fact is clear: all the men Anson brought to Canada were devout Christians — mainly Presbyterians — teetotalers, competent craftsmen, and dedicated workers. They were, in fact, steeped in the Phelps Dodge ethic.

The most prominent member of the group was Theodore W. Buck, who had been employed by Anson since 1866. When William Earl Dodge took over the presidency of the Georgian Bay Lumber Co. in 1873, Buck would be elevated to a senior management position, but when he arrived in 1871 he was general manager only of the three mills of Georgian Bay Lumber. William H. Russell was an experienced store manager. As assistant manager of the mills, his primary responsibility was to organize and manage the complex

Theodore W. Buck
1840-1881
– Huronia Museum

J.C. Else
– Huronia Museum

network of stores, warehouses,and farms. The stores of Georgian Bay Lumber, managed by local store-keepers responsible to Russell, were operated in Orillia, Port Severn, Sturgeon Bay, Waubaushene, and later Gravenhurst.

Stores for the Maganettewan and Parry Sound lumber companies operated independently in the early 1870s, sometimes purchasing supplies from Waubaushene. Later all the stores would be brought under the control of a subsidiary mercantile company. Principally, the stores provided supplies and equippage for the lumber camps and river drives and for the maintenance of the villagers. The company operated four farms in the early years — at Medonte township, Port Severn, Rama, and Waubaushene. The farms produced hay and oats for the draugh animals and provided pasturage for oxen in the summer.

John C. Else was a master millwright, responsible for installing and maintaining equipment in the enlarged mills at Waubaushene and Port Severn. Mill foremen Port Severn, Sturgeon Bay, and Waubaushene were, respectively, Andrew Meneilley, Henry Milne, and John R. Pierson. Phillip Schissler was yard foreman at Waubaushene, in charge of piling lumber under contract to the company. James H. Buck was manager of the Waubaushene store, and David J. Cooper was office manager.

These enterprising, religious zealots, most of whom were young men when they came to Canada, would make Waubaushene their permanent home. They raised families there; some of them are buried in the village cemetery. They and like-minded Canadians who joined the company in succeeding years would create the social climate that gave Waubaushene a reputation for sobriety and industry.

Anson took possession of the Waubaushene mill in July 1870 and the Port Severn and Sturgeon Bay mills in April 1871. When the mills ceased operating in the fall, an expansion program was immediately begun at Waubaushene and Port Severn. By the end of the summer of 1872, $50,003.36 had been spent on construction at Waubaushene and a further $33,268.77 at Port Severn. These figures included the cost of enlarging the mills, extending the docking and piling facilities, and building stores, warehouses, offices, and homes for married staff members. Consistent with the Phelps Dodge approach, schools and churches were built at both locations.

Roman Catholic churches were not characteristic of Dodge mill villages in the United States. The Dodges did, however, welcome

Roman Catholics in their Georgian Bay villages and allowed the diocese to operate churches. They really had no choice, because most of the bush workers on whom the logging operations depended were French Catholics who had followed the advancing lumber trade from Quebec through Glengarry County and up the Trent watershed to the Georgian Bay. Before the French came, Irish Catholic shantymen predominated. Names like Callaghan, Kelly, and Musgrove were common on company payrolls.

In 1878, 100 Roman Catholics were reported to be living in Port Severn, 100 in Muskoka Mills, 250 to 300 in Parry Sound and vicinity, and 200 in Byng Inlet.[12] Reference has already been made to the presence of a Catholic church at Waubaushene, when Anson bought the mill. Although the Christies were Presbyterian, they permitted a Catholic church at Port Severn — the lumbermen's church — erected in the mid-1860s for French-Canadian loggers. Dodge added Protestant churches in both communities.

It is not clear where the first Protestant church stood in Waubaushene, as the village was not divided into lots until 1894; thus the lot on which it stood cannot be traced. It was probably erected on or near the site of the present Union Church that was built as a replacement in 1881. The Port Severn church was built on the south side of the river, just west of the mill. A report in 1873 described the Waubaushene place of worship as "the neat little church . . . erected by the liberality of the lumbering company," while the Port Severn church was alleged to be "neat and comfortable."[13] The company not only built the churches but also provided coal oil for lighting and wood for heating, free of charge. Salaries of the student preachers were, however, the responsibility of church members. Dodge did not keep the Sturgeon Bay mill long enough to build a church, but services were held in the village, in private homes.

The three major Protestant denominations — Presbyterian, Methodist and Church of England (Anglican) — shared the church for Sunday services that were conducted at scheduled times. In the beginning, the Presbyterian congregation outnumbered the other two, mainly because the mill foremen and other American workers were Presbyterian and "warmly interested in the mission."[14]

The Presbyterian mission was organized in 1873 by the presbytery of Orillia under the direction of Reverend Mr J. Gray. The student missionary who ministered to the Waubaushene, Port Severn, and Sturgeon Bay congregations was Mr W. Frizzell of the

*The Lumbermen's
Church, Port Severn*
– Simcoe County
Archives

Mill Manager's House and Methodist Church, Port Severn c1890
– courtesy Robert Thiffault

Student Missionary Society of Knox College, Toronto. In August, Gray visited the mission and conducted services. It was reported that at Waubaushene "he dispensed the Lord's Supper to upwards of thirty communicants, and four children were baptised,"[15] while eight children were baptised at Port Severn. The Methodist and Anglican churches operated similar missions. Theodore Buck organized a Union Sunday School, open to any child or young person in the community who wished to attend. He, Dr John Hanly, and James Scott — all Presbyterians — were at various times superintendents of the school.

Most of the men who worked in William Hall's mill were single; consequently no school existed at Waubaushene. But since many of the men brought up from Pennsylvania by Georgian Bay Lumber and others hired in Canada had wives and families, a school, like the church, became an essential appurtenance to the village. The company built a fine little school, but it had no formal control over the running of it. In order to qualify for the per-pupil grant paid by the provincial Department of Education, it had to become a public school, with an elected board of three trustees. Thus, under the authority of the Act to Improve the Common and Grammar Schools of the Province of Ontario, passed in 1871, a school section — S.S. No. 12 Tay — was established and a board of trustees elected. Because the law stipulated that only British subjects could qualify as trustees, none of the American immigrants was eligible. The first three trustees were Andrew M. Adam, John Hanly, and Alexander McDonald. The first teacher was Irene Day Purkiss, who received an annual salary of $275.

In an ordinary school section, operating funds, not covered by government grants, were raised by a levy on all taxable property in the section, but since nearly all the pupils who attended the Waubaushene school were children of company employees living in unassessed company houses, the company paid the bulk of the operating expenses of the school, in lieu of taxes. Consequently, the Waubaushene school was better equipped and maintained than most village and rural schools and teachers' salaries were quite good. Similar schools and public school sections were established at Port Severn and Byng Inlet.

Although Waubaushene was a picturesque and comfortable village in 1872, it was isolated. Before the Midland Railway entered the village in 1875, the only means of reaching the outside world was by stage-coach, twenty-two miles to Orillia, where a connection could be made with the newly built extension of the Northern Railway, or by steamer to Collingwood, where one could board the main line of the same railway. The isolation worried the Pennsylvania women, especially because there was no ready access to medical services. Soon after they arrived, their plight was accentuated when a young local woman gave birth to a baby, without the assistance of a doctor. Although irrelevant to the welfare of the baby, the fact that the great-grandmother of the child was only forty-five years of age dramatized the event, raising anxieties even more. Unlike their more stoic and conforming husbands, the women were not willing to sacrifice security for the benefit of the lumber company. They threatened to return to Pennsylvania and leave the men to their own devices, unless a doctor was stationed permanently in Waubaushene.

Anson authorized the employment of young Dr John Hanly, who was persuaded to go to Waubaushene from his practice in Craighurst with the promise of a free house, free firewood, the keep of a horse, and $100-a-year retainer paid by the company. Hanly occasionally travelled to Byng Inlet, on the steamer *Magnetewan*, and regularly to Port Severn, to deliver babies ($5 each), set broken bones, and treat the sick. He became a pillar of Waubaushene society in the twenty-five years he doctored there. He was an elder and secretary of the Presbyterian church, superintendent of the Sunday school, and chairman of the school board for many years. He retired in 1896 because of failing health.

Anson discovered that his business "philosophy" — whether feigned or genuine — was compatible with traditional Canadian values and ideals. Here was a country that had been founded by political refugees who fled from his own country, to whom "republicanism" and "democracy" were wicked words, and whose descendants, guided by Loyalist traditions, had a brand new government based on the concept of "Peace, Order and Good Government." Here was a country where a handful of Scotsmen, with the same Spartan, Calvinist background as the Dodges, shaped society by controlling trade, banks, financial houses, major universities, and to a large extent the government. Here also was a country whose paternalistic government sought to protect its citizens from the evil of alcohol and the sin of sloth with stringent liquor laws. Surely in such a country, so rich in pine and other resources, whose social and entrepreneurial attitudes were so consistent with his own, where politcal leaders courted his favours and newspaper publishers extolled his virtues, he could not do other than succeed. He had carried out his side of the social contract, by providing for the social, moral, and intellectual welfare of his people. All that was required now was the co-operation and faithful discharge of duty by bush workers, managers, and mill hands and " A Deo victoria" would be assured.

Felling a pine tree   – National Archives PA 120333

Chapter Five

A Lumbering Year

In establishing a lumber industry in Byng Inlet, in 1868, Anson Dodge had to start from scratch, building the mill, surveying his timber limits, establishing depots, erecting shanties, and cutting cadge roads. But at the southern end of Georgian Bay, where the Georgian Bay Lumber Co. began to operate, a pattern of logging had already been established. Some bush shanties existed, and there was a body of experienced lumbermen available, who had either owned or worked for the existing mills. Anson hired many of these men and, as we have seen, gave shares in his company to the more experienced ones. Also, the men he brought from Pennsylvania, though unfamiliar with logging techniques in Canada, were experienced and competent mill operators and managers.

When Anson bought the three mills of the Georgian Bay Lumber Co. (1870-71), he purchased the lumber in stock and the logs that were either in stock at the mills or in the bush, ready for the spring drive. There was no break in the annual cycle of lumbering operations as a result of Dodge's takeover. As with the change of ownership of any industry, loggers and mill workers did not detect any immediate change in the daily routines of their lives, the one exception being that any worker who either made or purchased liquor had to desist or leave the company's employ. From the day Anson took over the mills, and for the next couple of decades, the consumption of alcohol on any of the company's premises was absolutely forbidden.

In this chapter we examine the seasonal cycle of operations, from the felling of the pine trees to the delivery of sawn lumber and squared timber to customers. The lumbering year 1871-72 has been chosen because that was the first full year (and the last) that Anson operated all the company's mills. The location is the north end of Simcoe County and the southern part of Muskoka, between the Musquash and Severn rivers. A similar sequence of operations was taking place at Anson's other mills, at Byng Inlet, Collingwood, Longford, and Parry Sound. And other Ontario lumber companies were duplicating the processes.

Forest technology in 1872 was essentially what it had been for the previous fifty years and would continue in the same fashion, with only minor changes, as long as the Georgian Bay Lumber Co. existed. Horses would replace oxen for hauling logs; steel cross-cut saws would replace axes for felling trees, speeding up harvesting somewhat; substantial camps would replace the crude shanties; and camp food would be improved. But invention of new equipment and techniques came slowly, so that logging changed little in the century between 1826, when lumbering began in earnest in Ontario, and 1920, when Georgian Bay Lumber closed down its last mill.

Sawmill technology, by contrast, changed considerably, especially in the second half of the nineteenth century. The insistent pressure of demand for lumber in North America fuelled efforts to increase the speed at which logs could be cut up. This was accomplished by a succession of inventions, beginning with the "gang-saw," that is, two saws and then many saws in the same frame (1835 onward), the circular saw and "gang-circular" (1840 onward), and the band saw and the "gang-band" (1890s).

Each new type of saw invented and improved cut so fast that the mills became choked up. Hence a whole set of inventions designed to feed and clear saws faster followed: jack ladders, edging saws, mechanical carriers for feeding logs into the saws, "niggers 'heads" or rams for turning logs over, tracks and trolleys for moving lumber to the piling yards. By the time Anson began operating in the 1870s, gang saws, which he had in all his mills, had reached technical perfection, and so, other than for the installation of band saws at the turn of the century and the addition of better-quality circular saws, milling operations of the Georgian Bay Lumber Co. changed very little over the period.

With the purchase of the three mills of Georgian Bay Lumber, Anson had acquired 202 square miles of timber limits on the Severn River and some 14,482 acres of land in fee simple scattered throughout Matchedash, Medonte, Orillia, Tay, and Tiny townships. He took over as well timber rights to 3,462 acres that Christie had purchased from settlers in the same townships. The first task in combining the operation of the three former separate mills was to hire a timber cruiser, or, as Anson's bookkeepers called him, an "explorer," to survey the company's properties to find the best stands of pine, so that depots and bush shanties could be located most advantageously.

The man hired was Patrick Murphy, who surveyed Matchedash, Medonte, Orillia, Tay, and Tiny townships. Murphy submitted detailed reports, listing the quality and quantity of pine and other timber, and the types of soil and land features, on each lot. On the basis of these reports, the company determined which lots to resell and which to harvest. Murphy also identified properties with good stands of pine belonging to settlers. If these lots were near company property, the timber was bought from the settlers; the deeds included permission to enter the properties for a specified number of years to remove the pine. Aubrey White was employed as timber cruiser in Muskoka. He also acted as an agent for the company, with authority to purchase standing timber from settlers in Muskoka.

Most of the better-quality pine remaining in Simcoe County was located in the north end of Orillia township and in Matchedash. Seven shanties were constructed there. There were other reasonably good stands of pine in Medonte township, near the headwaters of the Sturgeon River and in Tiny township, at the headwaters of the Wye River. Bush camps were established at these locations as well. Other camps were operated on the Severn River, on the shore of Georgian Bay near Honey Harbour, and on the shores of Lake Muskoka. There were, in addition to the camps operated by the company, several camps belonging to independent contractors, or "jobbers" to whom the company paid between $2.25 and $2.50 per thousand feet for cutting and skidding the logs. In *The Free Grant Lands of Canada, 1871*, McMurray reported that Dodge and Co. employed 800 men in Muskoka.[1] In the same year the Orillia *Northern Light* claimed that Dodge and Co. "employ more than 1,000 choppers"[2] in the woods operations between Byng Inlet and Muskoka.

The shanties were typical camboose shanties, characteristic of the period. These were crude log structures with a "camboose," or open fire, on the floor, centred under a great wooden chimney, about eight feet square, tapering upward to six. The fire, built on a clay or sand hearth and held in place by a square of logs, served for both cooking and heating. Two or three rows of bunks lined the walls. Generally the camboose kept the lumber-jacks in excellent health during the long, cold winters.

Georgian Bay Lumber's camps were a few years ahead of most other lumber camps in that they had separate rooms furnished with stoves for cooking. The camp contained also a store-house and a stable for horses and oxen. Each of the camps in Simcoe County housed from twenty to twenrty-five men, including the foreman, cook, scaler, and, later, a saw filer.

It cost from $2,000 to $2,500 to equip a camp. Personal property, such as pots, pans, dishes, lamps and lanterns, ticks and ticking, blankets, towelling, and wash basins cost about $300, depending on the number of men. Standard equipment for logging included axes (usually three or four dozen), at $16.50 per dozen, axe handles at thirty cents each, files, axe stones, adzes, chisels, augers, chains, ropes, and saws. Bob sleighs cost $30 each; a team of horses about $385; and a yoke of oxen, if available, only $225. Horse harness was worth $12 a set, whereas wooden ox yokes cost only $2 to $2.50.[3]

The company did not provide canthooks and peaveys. Each logger was expected to provide his own; indeed, he preferred it that way. The canthook was the logger's principal tool. With a medium-length wooden handle, fitted with iron jaws at the lower end, and equipped with an adjustable iron hook, the canthook was really a lever for moving logs. Used properly, a canthook could roll a heavy log, change the direction of roll, or stop a rolling log. It was used for piling skidways, loading and unloading sleighs, and breaking up log-jams. Its proper and quick application could prevent a logger from sustaining a broken leg or even losing his life. For this reason, the lumber-jack, like the western gun-fighter who relied on his own six-shooter, preferred his own canthook with a weight and balance familiar to him, enabling him to use it automatically and effectively in any emergency. The peavey (named after its inventor, Joseph Peavey) had a longer handle than the canthook and was fitted with a metal spike at the lower end instead of jaws. It was used on log drives.

Food and supplies were hauled to camps regularly, either from a central depot, as with the Maganettewan Lumber Co.,or from stores (at Orillia, Port Severn, and Waubaushene), as with the Georgian Bay Lumber Co. Supplies for the camps near Washago and at Ragged Rapids (now Hydro Glen) were taken from Orillia in the

Inside a "Camboose" Lumber Shanty   – Ontario Archives ACC10010-44

Lumberjacks on the Severn River   – National Archives C-11738

autumn by barge or on the steamer *Camilla* and either hauled to the camps by team or freighted down the Severn River in scows. Scows also carried equipment and supplies from the stores at Waubaushene and Port Severn to camps near the Severn, Wye, Sturgeon, North, and Black (now Matchedash) rivers in September and October. In the winter months, supplies were hauled by team, usually under contract, along cadge roads to the camps. Hay for the horses was bought from local settlers. Epplett's mill at Coldwater supplied oats at fifty cents per bushel for the camps in Orillia and Medonte townships. Oats and hay were also brought to Waubaushene and Port Severn by schooner from more distant points.

Lumber-jacks in Georgian Bay Lumber camps ate extremely well, probably better than workers in other company camps at that time. The staple of diet was, as in all other logging camps, salt pork from Chicago ($18 per barrel), beans ($5 per 100 pounds), molasses (eighty cents per gallon), and potatoes ($1.25 per bushel). Tea cost eighty-five cents a pound. Flour for bread, biscuits, and pies cost $7 per barrel. Apple or currant pies were served on Sundays. Syrup ($32.50 per barrel) poured over bread was a special dessert, frequently provided during log drives. Cooks became quite adept at relieving the monotonous taste of pork and beans by flavouring them on occasion with hops. Mustard, cloves, and cinnamon were also commonly used for flavouring.

The men had to provide their own tobacco. Most of them chewed, because smoking was not permitted in the mills or piling yards. The company charged fifty cents per pound for both chewing and smoking tobacco.

A camp of twenty men consumed, on average, in one week 1½ barrels of pork, 2½ barrels of flour, fifty pounds of beans, two bushels of potatoes, and five pounds of tea. Two teams of horses ate three bales of hay and four bushels of oats per week. The men were given free board and paid from $15 to $18 per month. One typical camp — Thomas Rawson's at Black River — required $1,401.11 for supplies over the seven-month logging season; wages totaled $1,839.06, making a total cost of $3,240.19 for operating the camp. This figure does not include horses or equipment, most of which was returned to the stores and used again the following year. At the end of the season, horses were boarded in a stable in Orillia or housed in the company's own stables at Port Severn and Waubaushene.

The annual cycle of logging began in mid-September, when the camp foremen and skeleton crews began stocking the camps and cutting roads. Occasionally new shanties had to be built. By mid-October, when the mills closed down and the harvest was finished, the company's mill hands and local farmers looking for winter employment trekked into the camps. The loggers' day started at 4 a.m. After a breakfast of beans and bread, they walked up to three miles through the bush to where the cut was taking place and, in the cold months, waited beside a bonfire until it was light enough to start work.

After the invention of the steel cross-cut saw, work gangs had three members, but in 1872 loggers worked in gangs of five: two choppers felled the trees and limbed them; two sawyers sawed the trunks into logs 12, 14, or 16 feet in length, and a skidder hauled the logs to a central clearing and piled them on log skids in huge piles, called skidways or rollways. Others were engaged in cutting draw roads which led like veins from the skidways through gulleys and valleys into the artery of the main road leading to a dump at the edge of a stream, river, or lake. Still other men were engaged in building and repairing dams and timber slides, where necessary. At eleven o'clock a hot lunch of pork, bread, and tea was brought out from the camp. Loggers returned to the camp at dark, and supper was served at six o'clock. Sunday was a day of rest.

The government required every camp to employ a trained scaler to count the logs and measure the cut each day. Cumulative records were kept. Because the timber dues of $1.50 per thousand feet board measure paid to the government were based on these measurements, government inspectors visited regularly to ensure the records were accurate. Pine trees less than thirteen inches in diameter were ignored, the average diameter of trees felled being about fifteen inches. Large trees, three or four feet in diameter, were set aside for boom timbers, which were needed for towing logs on Georgian Bay, especially in rough weather. A twelve-foot log fifteen inches in diameter translated into 142 board feet of lumber.

The average rate of production of a logging gang was sixty logs per day, about three-quarters of which were skidded. The winter's cut of a camp naturally depended on the number of men employed. In the winter of 1873, Rawson's Black River camp produced 6,340 logs which measured 1,035,859 board feet. (Logging camps on the Magnetewan River were larger than those in

Simcoe County. Scalers' reports for 1872 record a total cut of 70,464 logs, or 10,269,171 board feet, for the six camps operating there. Two of the largest camps produced about 2¼ million feet, with the average being about 1½ million.)

In the season of 1872-73, the eleven or twelve shanties of Georgian Bay Lumber produced slightly more than 10 million board feet of logs. Jobbers in Simcoe County delivered a further 6,606 logs, measuring 1,290,360 board feet, to the company's log dumps, while a host of other jobbers in Muskoka and Victoria counties (Longford mill figures were included with Georgian Bay Lumber's calculations in 1872) produced 13,631,000 board feet. Some of these jobbers, of whom there were 118 altogether, were local settlers who delivered no more than thirty or forty logs, but a few operating large camps delivered as many as 3,700 logs. Counting logs in inventory from the year before, Georgian Bay Lumber's total production of logs that year was about 25,374,934 feet. This was a considerable reduction from the year before, when the company took out 32,972,901 feet of logs.

Foremen tried to get the winter's cut of logs on skids by early January, so that all the men could be employed in hauling logs to the water's edge for the spring drive. There was greater urgency in Simcoe County, where spring set in earlier, than further north in Parry Sound district, but even there log hauling usually started about the first of February. The log-cutting crews of autumn then became loading gangs, and teamsters who skidded logs from tree stumps to skidways drove the teams that hauled logs to the dumps.

The number of sleighs used in a winter's haul depended on the number of logs on skids and the length of the haul to the water's edge; eight seems to have been the normal number of sleighs used in Georgian Bay Lumber's larger camps. The logs were piled in dumps on the banks of rivers, streams, and lakes. When the ice broke up in April, the logs were rolled into the water to be driven free and loose down the swollen rivers until they reached either Georgian Bay or Gloucester Pool, where they were boomed and towed to the mills. Because other firms ran logs down the same rivers, each log was hammer-stamped with the company's mark[4] for identification purposes.

Log drives, especially on the Severn and Musquash rivers, were extremely arduous and dangerous. The rampaging rivers, swollen by melting snow and spring rain, propelled the logs along

Skidway   – Ontario Archives S12103

Log dump   – Ontario Archives S12144

River drivers — National Archives 22454

Running logs in Muskoka — National Archives C25720

at great speed, through dams, along slides, and through chutes. Drivers had to follow the logs all day and part of the night, pushing stuck logs off the banks and shallow places and occasionally breaking up log-jams, the most difficult and dangerous task of all.

The North, Sturgeon, and Wye rivers, although discharging more water then than they do now, were relatively gentle-flowing streams. The problem in these rivers was to store enough water behind dams to permit flushing the logs to the rivers' mouths before the spring freshet subsided. The drives on these rivers were never large and ended in the mid-to late 1870s. The Black, or Matchedash River, fed by two lakes in the southwest corner of Matchedash township, generally had enough water for the spring drive; shallow rapids created the major problems there. To facilitate the movement of logs and to prevent jams from forming, the company built eleven dams and seven slides on the Black River and its tributary streams.

The Severn and Musquash — the Severn fed by Lake Simcoe and Lake Couchiching, and the Musquash by the Muskoka Lakes system — are large rivers. They were the principal water courses down which logs were floated out of Muskoka by Georgian Bay Lumber for the next twenty-five years. Rarely was an insufficient supply of water in these rivers a problem. Logs dumped into the Severn River above Sparrow Lake had to be warped across the lake, but in the twelve-mile stretch of river between Macdonald's Rapids and Little Chute, the spring current kept the logs moving without assistance.

The main obstacles on this stretch were Ragged Rapids and Big Chute. The falls in the river at these two spots were respectively twenty-nine and fifty-eight feet in just a few hundred yards. (Not surprisingly, hydroelectric-power plants were later built at these two falls: one by the town of Orillia at Ragged Rapids in 1898, the other at Big Chute by the Simcoe Railway and Power Company of Midland in 1910.) Georgian Bay Lumber built dams and slides at both rapids, and some blasting of the rock in the confined channel was done at Big Chute, but despite these improvements, massive log-jams occurred frequently, especially in late summer, when the water-supply was lowest. The company's river drivers frequently avoided Big Chute altogether by running logs down Pretty Channel into Lost Channel and Six Mile Lake and through the Six Mile Lake dam and slide back into the main river again, just above Little Chute. In 1872, J.C. Hughson, with whom Dodge had a partnership in the mill at Muskoka Mills, ran the company's logs down the

Musquash River from Lake Muskoka and billed the Georgian Bay Lumber Co. accordingly.

A catch boom — large logs chained end to end — was stretched across the river below Little Chute, where bags of 2,000 to 2,500 logs were collected. Similar booms were strung across the mouths of the other rivers. Logs floated down the Severn and Black rivers, destined for the Waubaushene mill, and other companies' logs bypassed the Port Severn mill through a dam and slide at the north end of the main island. There was a sorting jack below Port Severn, where skilful and sharp-eyed rivermen stood on the boom timbers, spearing the logs with pike poles and guiding them into the separate booms of each owner.

In 1883, Georgian Bay Lumber built and launched a steam tug in the river above Port Severn to haul booms of logs down Gloucester Pool and MacLean Lake (the entrance to Black River). After 1890, "alligators" were used regularly on Six Mile Lake and other quiet, inland lakes to move logs. These short, flat-bottomed, wood-burning, paddle-wheelers were equipped with windlasses capable of holding a mile and a quarter of five-eighths-inch cable. These amazing little craft could tow a small boom of logs from one end of a lake to the other, or when the boat was snubbed to the shore, the windlass could warp a large boom of many thousands of logs weighing tens of thousands of tons. The alligator could also winch itself across portages to catch up with the log drive in the next body of water.

But in 1872, the only way that booms of logs could be moved across quiet, open stretches of water was to warp them with capstan cribs. A capstan crib was a raft of squared pine timber, in the centre of which was a spool three or feet in diameter, fashioned out of laminated planks. The spool revolved around an iron post, rooted in the floor of the crib. An eight-foot arm, protruding from one side of the capstan, allowed four or five men to wind up a manila rope, which was fastened to a boom of logs a quarter-mile away. To prevent the crib from being warped back toward the heavier boom, the crib was anchored to the bottom with a four-or five-hundred-pound anchor, or it was snubbed to a tree or rock on a convenient point of land or island. In this way, booms containing 2,000 to 2,500 logs were warped in quarter-mile strides over the eight miles of river between Little Chute and Port Severn. Some companies used horses to turn the capstans, but whether the Georgian Bay Lumber Co. did so or relied solely on manpower is not known.

Log slide, Muskoka   – Ontario Archives S3625

Log jam, Big Chute 1896   – courtesy C.A. Stocking

Logs were released into the rivers at regular intervals, so that there was always a steady flow of logs descending the rivers and arriving at the mills. The more distant logs were generally driven down first. The river drives were finished in mid-to late August, just before the logging cycle began again. Logging lore is replete with tales of drinking sprees that took place when the hard-driving, hard-drinking rivermen were paid off at the end of the log drives. But such spectacles were not typical of Georgian Bay Lumber's river drivers — certainly not in Port Severn or Waubaushene.

Logs from the Wye and Sturgeon river drives supplied the small Sturgeon Bay mill. Excess logs from these drives and all of the North River drive were towed to the mill at Waubaushene. Logs of the Musquash River drive not consumed by the mills at Muskoka Mills, and logs from the Severn and Black River drives not cut at Port Severn, were also towed by tug to Waubaushene.

Through the years, Georgian Bay Lumber would own a number of steam tugs of various sizes and horse-power, but in 1872, it had three coal-fired tugs on Georgian Bay: *Lilly Kerr* (acquired with the purchase of the Port Severn mill), *Prince Alfred* (in which the company owned a two-thirds interest), and the *Mittie Grew*. The Collingwood mill had the *George Watson;* the Parry Sound Lumber Co. the *Wave*; and the Maganettewan Lumber Co. the *Minnie Hall*. The name of the latter tug suggests that Anson may have acquired her with the purchase of the Waubaushene mill from Hall and later transferred her to Byng Inlet.

All the tugs were paddle-wheelers. *Lilly Kerr* was built for A.R. Christie at Lock 5 on the Welland Canal in the winter of 1868-69, at a cost of about $8,000. She was 81 feet long, with a 16-foot beam. Her side paddle-wheels were seven feet in diameter. Her engine's cylinder had a 20-inch bore with a 22-inch stroke. Her boiler was 78 inches in diameter and 12½ feet long, and her smoke stack was 20 inches in diameter.

The smaller tug, *Prince Alfred*, was built at Brockville in 1867 and enlarged at Port Severn. She was 61 feet long, with a 13-foot beam; her paddle-wheels were 4½ feet in diameter.

Mittie Grew, named after Anson's wife, Rebecca (Mittie was her nick-name), was Anson's personal vessel. He had her splendidly adorned with coco matting, deck chairs, and cushions with costly enameller cloth covers. Because Anson used her primarily for pleasure or business, *Mittie Grew* sustained an operating deficit. The deficit in 1871-1872 was $870.71.

The other tugs were work boats. Daily, they chugged back and forth between Waubaushene, Sturgeon Bay, Port Severn, and the rivers' mouths, towing booms of logs to the mills; they towed schooners and barges into and away from the loading wharfs, and they carried freight from one company establishment to another.

Generally, in April or early May, as soon as the ice was cleared from the log ponds, the mills began sawing, there usually being enough logs left over from the previous year to get the mills started before the arrival of logs, in steady procession, from the current river drives. In the spring of 1872, the Anson and Parry Sound mills began cutting in April, the Sturgeon Bay, Longford, Page, and Collingwood mills in mid-May. The Port Severn and Waubaushene mills did not begin to operate until 1 June, probably because the expansion and renovations carried out at both mills delayed the opening date.

The Waubaushene mill was similar in design and operation to all the other large lumber mills in Ontario, indeed, in all of North America at that time, with one possible difference: Theodore Buck's obsession with order, efficiency, and neatness ensured that the mill was more attractive and better managed than most. One journalist for the *Canada Lumberman* called it "a model sawmill."[5] No photograph of it survives, but a sketch on company letterhead and a detailed inventory of equipment made by Theodore Buck in 1873 give us a fair idea of what the mill looked like and how it operated.

The main building, 112 feet by 65, feet was two storeys high. The lower level housed the engines and various pieces of machinery for working the saws and rolling platforms on the second floor. On a level with the second floor was an extensive platform which, supported by trestle-work, carried the several miles of elevated tracks out to the lumber yards. The building was painted grey, and the extensive trestles and other appurtenances were painted white. The store and numerous warehouses were erected nearby, and a boat-house protected the tugs. The area surrounding the mill was raked clear of all debris, creating an appearance of neatness and orderliness.

The saws in the second storey consisted of two stock gang saws, one 52 inches wide, the other 36 inches; one edger with four saws; one stock trimmer; two slabbing saws, each 36 inches wide; and one circular saw, 66 inches in diameter. There was also a lath mill, a planer, and a vertical drill. The whole operation was mechanized. Two jack-ladders hauled sawlogs in endless procession from the water into the mill, where they were rolled onto a log car, one at a time. This log car — called a "snake" — moved the log forward between two circular saws, or slabbers, adjusted to an appropriate width. The slabbers removed the bark and rough wood from two opposite sides of the logs. After the logs had passed through the slabbers, they were rolled onto a framework bearing a number of iron rollers that carried them toward the gang saw. At an appropriate point a steam ram, or "nigger head," came up through the floor and rolled the heavy logs, flat sides up and down, onto another car that moved them slowly endwise into the upright gang saws.

The upright gang saw worked on the same principle as the ancient whip-saw, the main difference being that steam-power pushed a number of saws up and down instead of two men pushing one. The parallel saws in the gang — from forty-eight to sixty in number — could be adjusted for any desired width of board. The gang saw could handle a log as large as five feet in diameter, but generally two to four smaller logs were fed through at once.

When the several rough boards emerged simultaneously from the gang saw, they were removed to another car on which they were passed sideways through a circular saw, or trimmer, that cut them into desired lengths and trimmed off the rough ends. The finished boards were then piled on tramcars and rolled along the elevated tracks to the lumber yard. Lumber piling at Waubaushene and Port Severn was done under contract, because that method was found to be cheaper than employing the company's own men. The contractor at Waubaushene was Phillip Schissler.

Slabs were taken to the lath mill and cut into four-foot laths and tied in bundles of one hundred. The lath mills at Waubaushene and Port Severn were operated under contract, the company paying the contractors seventy-five cents per thousand pieces for lath, cut and bundled.

Long, shallow troughs reached around the building; into them sawdust, bark, and refuse wood were dumped. Endless chains with traverse pieces of wood attached scraped all the refuse out of the building and either into the boiler room or away to a large pit, where it was burned. (The tall bottle-shaped burner for disposing of waste seen in many photographs of Waubaushene was not constructed until later in the decade and perhaps not even until a new mill was built in 1881-82.)

The motive power consisted of four steam-engines, generating 90, 40, 80, and five horse-power. The boilers were 29 feet long and 50 inches in diameter. A network of long 24-inch-wide leather belts led from the drive shafts to work the saws, log cars, and other bits of machinery on the second floor. To maintain their flexibility and to prevent them from becoming friction-glazed from the pulleys, the belts were constantly lubricated with oil. Belt oil was very expensive — forty-two cents a gallon in 1872. Hundreds of gallons were used in a season's operation.

The whole operation, from the loading of a log onto the jack-ladder until the sawn boards were taken to the piling yard, lasted only a few minutes. Both gang saws worked simultaneously, and the mill operated from six o'clock in the morning until six o'clock at night. Despite Anson's avowed commitment to Christian principles, he insisted that the mill operate on Sundays, a decision that would not have endeared him to the devout Theodore Buck. The mill had a cutting capacity of about 100,000 board feet of lumber per day, 80 — 85,000 board feet per day being the norm in 1872. But the saws could cut much more than that if pushed. Mill records show that on 22 June 1871 — the peak year before the slump in lumber sales — 147 hands cut and piled 153,873 board feet of one-inch and one-and-a-half inch lumber.

In 1872, the Waubaushene mill employed eighty-seven men and eighteen horses. On average five men worked on the logs, feeding the jack-ladders; fifty to sixty worked in the main mill; and five or six were engaged in trimming. The piling contractor employed another five or six men in the yard, and some were usually employed loading vessels. Well over one hundred men were employed in the village altogether. Workers were shifted from one job to another as the need arose — one day loading vessels (later box cars), another day on construction or general repairs.

Wages averaged $26 to $35 per month. Married men were given free housing, and single men were provided accommodation in boarding-houses, but they had to pay board. In 1873, there were two boarding-houses in Waubaushene, accommodating sixty men, and twenty-six houses, accommodating thirty-six families. To be ensured of a free house, the employee had to work in the lumber camps in the winter, unless he was a technician or manager whose services were required around the village all year.

The Port Severn mill had a similar number of saws and employed about the same number of men, so that its cutting capacity was about the same, perhaps even a little higher, than the Waubaushene mill. The main difference was that the Port Severn mill was powered by water, using iron turbine wheels, one for each saw. Because there was no steam plant to consume waste, and until a burner was constructed, most of the refuse was loaded into scows and taken to an island in Georgian Bay to be burned; a good deal of it was simply dumped into the Bay.

It is not known what kind of saws existed in the Sturgeon Bay and Longford mills, but, in 1872, the cutting capacity of both mills was quite small — around two million board feet per season.

Each of the mill managers made daily summaries of the amount and type of lumber cut, the number of men employed on each operation, and the daily wages paid. At the end of the month — using wage summaries and other mill expenses, such as the amount of lubricating oil consumed, the cost of belts replaced, and files used — the managers calculated an average cost of manufacturing one thousand board feet of lumber. In the summer of 1872, the average cost of manufacturing lumber at the Waubaushene mill was $1.52 per one thousand board feet, at Port Severn $1.28, and at Sturgeon Bay $1, making an average manufacturing cost of $1.36 per one thousand board feet at the three mills of Georgian Bay Lumber. Similar figures are not available for the other Canadian mills owned by Anson, but a surviving mill report of the Parry Sound Lumber Co. for one week in June 1872 shows an average cost of $1.30 for manufacturing one thousand board feet.

By 1 October 1872, the Port Severn mill had sawn 8,981,557 board feet of lumber and the Waubaushene mill 8,535,266 feet; the smaller Sturgeon Bay and Longford mills had cut 1,755,181 and 2,242,354 feet, respectively, making a total season's cut of 21,514,358 board feet at the four mills. The Anson, Collingwood, Page, and Parry Sound mills cut, in total, another 27,961,112 board feet, making a total cut of 49,475,460 board feet at all the mills. With 13,659,655 board feet of lumber in stock at the mills from 1870 — 71, Anson had 63,135,115 board feet of lumber to dispose of in the 1871 — 72 season. (When the mills stopped operating, at the end of October, they had cut another 3 or 4 million board feet,

The Parry Sound Mill 1876　　　　　– Parry Sound Public Library

Squaring timber　　　　　– Ontario Archives

bringing the amount closer to 68 million feet.) This was, indeed, lumbering on a large scale.

The other branch of the industry — the square-timber trade — was a new experience for Anson when he came to Canada. Timber making, as the process was called, was a uniquely Canadian enterprise which had its origin in the Napoleonic Wars. As a maritime trading nation, Britain was completely dependent on its merchant ships and naval forces. Since ships were made of wood — the hulls of oak, the masts and spars of white pine — Britain could not hope to remain a first-class power without an assured supply of oak timber and pine masts, in times of peace as well as war. With the disappearance of native English wood after 1500, Britain depended increasingly on foreign timber. From the end of the fifteenth century to the beginning of the nineteenth, Britain had imported timber for naval and domestic needs from countries surrounding the Baltic Sea. But those sources were cut off during the Napoleonic Wars, and Britain was obliged to turn to the colonies in North America to satisfy its timber requirements, especially for pine masts.

As a sizeable investment in capital was needed for cutting the large pine trees and transporting them to England, the trade could not be carried on profitably in masts alone, and so the navy had encouraged development of a general timber trade with the colonies. With cessation of hostilities in Europe, there was a strong possibility that British shipbuilders would once more look to the Baltic for cheaper and more readily accessible timber, to the disadvantage of Canadian producers. To protect the colonial trade from European competition, Britain erected a tariff wall, through which Canadian timber passed duty-free, from 1821 to 1842. Consequently, the square-timber trade flourished.

Because logs had to be transported across the ocean in ships, it made sense to strip away as much useless material from the trunks as possible. To facilitate piling in the holds of the ships and to make maximum use of the space available, the practice of squaring the logs into timbers evolved. To prevent accumulation of masses of waste material at harbours, timber manufacturers had the logs squared in the bush by craftsmen who possessed special skills not required of ordinary shanty men.[6]

The square-timber trade waxed and waned with changes in market conditions in Britain, where the timber was sold on commission by British agents and brokers, to whom it was consigned by Canadian shippers. The timber was sold at public auction and was generally bought by the large sawmill owners, who held a virtual monopoly over the disposal of timber in Britain. Millers' surplus timber was later sold in small quantities to builders and owners of small village sawmills.

Repeal of the Corn Laws in 1842 removed the tariff protection Canadian timber enjoyed, and British markets were reopened to Baltic timber. This dealt a severe, but temporary, blow to the Canadian square-timber trade. With the increase in population and the growth of the British economy caused by the Industrial Revolution, demand for Canadian timber soon picked up again. The period 1845-63 is considered the high-point for the export of Canadian square timber — 33 million cubic feet was shipped to England in 1863. The volume of trade moved up and down during the rest of the century, but always along a gradual downward curve in the volume of production. The amount of square timber produced in the twentieth century has been insignificant, especially after 1925.

In 1879 Ontario's commissioner of crown lands expressed concern about the wastage in the timber-making process and the resulting loss of revenue to the province. He estimated that one-fourth part of every tree cut down to be made into square or waney[7] timber was wasted, and provincial revenue suffered proportionately. The outer wood beaten off in squaring a log was the prime part of the tree, from which the best class of clear sawn lumber was obtained. Moreover, the upper part of the tree, from which a quantity of good second-grade lumber could have been produced, was discarded; it was left in the bush, and along with the fine wood beaten off, kindled forest fires that annually destroyed more timber than was cut down for commercial purposes.

The commissioner calculated the revenue lost to lumbermen and the province in the ten years after Confederation. Based on total known production — 119,250,420 cubic feet of square-timber — he estimated a loss through wastage of 477 million board feet of lumber. Using average market prices — $10 per thousand board feet for quality-lumber estimated to have been lost and $7 per thousand board feet for second-grade lumber wasted — he calculated that Ontario lumbermen lost $3,577,500 from squaring trees instead of sawing them. The corresponding loss to the provincial government was estimated to be $262,500. The loss of timber from forest fires fed by debris left from the trade was incalculable. The commissioner tried to discourage the square-timber trade, but profits from it were generally high, and so it continued for a few more decades.

The commissioner was equally concerned about the undesirable sociological impact of timber making on the province, in contrast to the positive contribution of sawmilling. Sawmilling was a factor in settlement, a high priority with the provincial government. Many sawmill employees, because of "their steady habits and value as workers"[8] were kept in permanent employment, summer and winter, and as a consequence took up lands near the mills. These lands were improved by the workers' families, and, as a result, sizeable areas of the province near the mills had become settled and cleared. (Waubaushene and Port Severn were typical examples.)

Timber makers, in contrast, were, by and large, without fixed homes or continuous employment. In the interim between termination of their engagements in the spring and their reengagement the following winter, they "too frequently remain[ed] idle and spent their earnings in a reckless manner, and [were] penniless and often in debt when they return[ed] to the woods."[9] Timber makers were definitely not God's workers.

The Georgian Bay Lumber Co. was never a large producer of square timber. On those occasions when it was deemed appropriate to make square timber, the work was always contracted out.

Shipping lumber from Georgian Bay was difficult in 1872, but an even greater task was finding reliable markets for it. As mentioned in chapter 2, Anson entered into an agreement with White and Co. late in 1872, by which White would take, at a fixed price, all the lumber Anson could produce. Before that, he sold lumber either directly or through forwarding agents to a host of US wholesalers: Chambers and Co. of Cleveland; Pritchard and Co. of Tonawanda; Ross and Co. of Detroit; D.L. Conch, of Oswego; R.A. Loveland and Co. of Chicago; and Mixer and Smith, of Buffalo. He shipped some of the poorer grades to his own yard in Philadelphia, and a quantity of lumber was sold at the mills to local customers and one or two Canadian dealers — J. Clements, of Toronto, and S. Hadley, of Chatham, being the two most prominent. Demand for lumber in both Canada and the United States was trending downward in 1872: by the end of September Anson sold only 37,254,860 board feet of the nearly 68 million feet he had available.

Before the Toronto, Simcoe and Muskoka Junction Railway reached Longford in 1873, lumber from the mill was towed across

Schooner at Waubaushene dock c1890   – courtesy C.A. Stocking

Loading lumber on a schooner, Waubaushene c1890

Lake Simcoe in a barge — *Couchiching, Morning, Isaac,* or *L. Bacon* — to Belle Ewart or Beaverton. From Belle Ewart, it was shipped on the Northern to Toronto, to the forwarding firm of F.P.G. Taylor and Co. Taylor shipped the lumber to American dealers, mainly in Oswego. The rail freight rate from Belle Ewart to Toronto was $2 per thousand board feet; the water rate from Toronto to Oswego was $1.50 per thousand feet. The forwarder charged 12.5 cents per thousand board feet for handling. There were, in addition, the cost of towing on Lake Simcoe and of loading and unloading rail cars — about $1 per car load of 7,000 to 8,000 board feet.

Lumber shipped through Beaverton on the Midland Railway of Canada was handled by the forwarding firm of Irwin and Boyd, of Port Hope. Rail freight by this route was also $2 per thousand board feet.

Shipping costs of lumber sent to American buyers from Waubaushene and Port Severn, whether through Taylor and Co. of Toronto or directly from the mills all the way by water, were about the same. Lumber handled by Taylor and Co. went through Collingwood. Costs per thousand board feet by this route were as follows: barge or schooner from Waubaushene, Port Severn, or Sturgeon Bay to Collingwood, $1.125; rail freight from Collingwood to Toronto, $2.20; schooner freight from Toronto to Oswego, $1.50; and forwarding charges, 12.5 cents; making a total of $4.95. The freight on lumber shipped from Waubaushene to Oswego by schooner was $5 per thousand board feet. The bulk of the lumber shipped from Georgian Bay mills to Amercan buyers went directly by water. Anson used three system of transportation. First, some lumber was carried by independent ship owners — the most expensive way. Typical rates per thousand board feet paid to the Independent shippers were: Byng Inlet to Chicago, $6; Waubaushene or Port Severn to Buffalo, $4.50; and Sturgeon Bay to Chatham, $5.

A second, cheaper method of shipping was by contract with Isaac May, Anson's neighbour at Roche's Point, who owned four barges on Georgian Bay: the steam barge *Isaac May* and three tow barges, *Muskoka, Waubaushene,* and *Port Severn.* Anson had invested in May's barges, hence the names. These barges were built especially for carrying lumber and could hold a good deal more than schooners. *Isaac May* was built in Welland, Ontario. She was 170 feet long, with a 30-foot beam and a 12½-foot hold. Her engine had a 30-inch cylinder to drive a single paddle, 9½ feet in diameter, mounted under a 12-foot fan-tail. She was also schooner-rigged.

She carried 500,000 board feet of lumber. The other three barges were all schooner-rigged. Each was 155 feet long, with a 30-foot beam and a 12-foot hold, and carried 525,000 board feet of lumber. *Port Severn* was built at Welland; the other two were constructed at Chatham, Ontario. The contract rate on May's barges for lumber shipped from Waubuashene to Buffalo was $3.50 per thousand board feet.

The third and cheapest means of shipping lumber was in the company's own schooners, of which it operated four. The flagship of the fleet was *Thomas C. Street*, owned outright by the company. Built in 1869, in St Catharines, Ontario, at a cost of $16,000, she was 139 feet long, had three masts and a twenty-three-foot-four-inch beam, and weighed 326 tons. She carried a crew of ten. Her master was George Morden. She could carry 330,000 board feet of lumber or 20,360 bushels of grain. On every trip that she and the other company schooners made to American ports, they brought back pork, coal, oats, and other supplies and equipment for the company's stores. *Thomas C. Street* was wintered in Toronto, at the Church Street wharfs, for which the company paid a moorage fee of twenty dollars for the season.

The other schooners operated by the company were *Kenosha*, Capt. Stephen Waggoner, master; *Queen of the North* , Capt. Peter Thompson, master; and a 72-ton schooner from Toronto, with the imposing name of *Dauntless*. *Kenosha* and *Dauntless* were operated probably under charter, but the company owned *Queen of the North* outright. She was built in the Nottawasaga River in 1861, originally as a brig, but in 1872 was rebuilt as a three-masted schooner, at a cost of about $11,000. She was 125 feet long, with a twenty-three-foot-two-inch beam, and weighed 293 tons. She could carry 14,000 bushels of grain or 260,000 board feet of lumber. The 292-ton *Kenosha* could accommodate 280,000 board feet of lumber, but the little *Dauntless* was hard-pressed to carry more than 80,000 feet.

Lumber was shipped "free on board" (f.o.b.) at the mills. On lumber shipped in its own schooners, the company charged the normal freight rates, but shipping did not cost nearly that much. For example, on one delivery of 273,594 board feet of lumber from Port Severn to Mixer and Smith at Buffalo, the captain collected from Mixer and Smith the standard rate of $4.50 per thousand board feet, or $1,231.17 for the load. The actual cost, including the cook's and crew's wages, provisions, port clearance, tug fees, towing charges

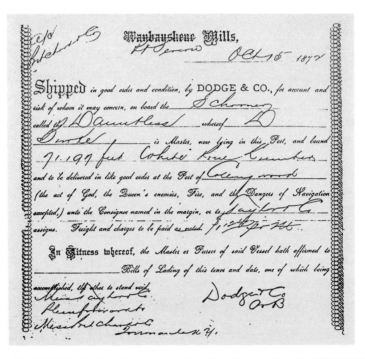

Waubaushene Mills,

Oct 15 1872

Shipped in good order and condition, by DODGE & CO., for account and risk of whom it may concern, on board the *Schooner* called the *Dauntless* whereof ____ *Dodge* ____ is Master, now lying in this Port, and bound 71.199 feet *White Pine Lumber* and to be delivered in like good order at the Port of *Collingwood* (the act of God, the Queen's enemies, Fire, and the Dangers of Navigation excepted,) unto the Consignee named in the margin, or to *Taylor & Co* assigns. Freight and charges to be paid as ____

In Witness whereof, the Master or Purser of said Vessel hath affirmed to ____ Bills of Lading of this tenor and date, one of which being accomplished, the other to stand void.

Dodge & Co
O. B.

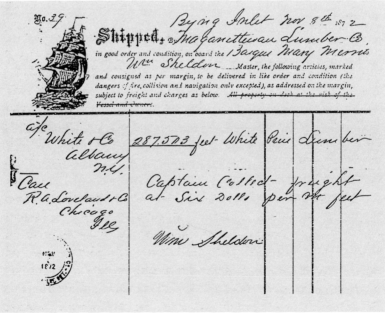

No. 29

Shipped, by *Ino Ganetteau Lumber Co*

Byng Inlet Nov 8th 1872

in good order and condition, on board the *Barque Mary Merritt* *Wm Sheldon* Master, the following articles, marked and consigned as per margin, to be delivered in like order and condition (the dangers of fire, collision and navigation only excepted), as addressed on the margin, subject to freight and charges as below. All property on deck at the risk of the Vessel and Owners.

a/c White & Co Albany N.Y.	287,573 feet White Pine Lumber			
Care R. A. Loveland & Co Chicago Ill	Captain Collect freight at Six Dolls per ____ feet			
	Wm Sheldon			

on the Erie Canal, sail repairs, and sundries, was $462.92. Thus, the company made $766.25 on that trip. In the financial year 1871-72, the schooners' accounts all showed operating gains on the season's trading: *Thomas C. Street*, $4,619.41; *Kenosha*, $3,411.43; *Queen of the North*, $747.44; and *Dauntless*, $358.39.

Accounts for each operation were kept in separate folios, one each for milling, stores, farms, stables, schooners, tugs, logging camps, river drives, lumber sales, and so on. Monthly balances for each account were entered in a master account for each mill and for the woods office in Orillia. Trial balances were taken off the master accounts at the end of each month and sent to head office in Barrie, where grand totals were compiled. These were transferred to the books of Dodge and Co. in New York.

Anson Dodge had hoped that the books would show a profit, but it soon became apparent that income from the Canadian mills was not nearly enough to meet payments on the large debts that Dodge and Co. had incurred in acquiring the mills. Georgian Bay Lumber was not contributing to "a Deo victoria," if God's victory was synonymous with making Anson the "wealthiest man in North America."

There is no question that the workers gave the "cooperation" that Anson's corporate dream envisioned — the mills were run efficiently, and the logging operations were cost-effective. Under Buck's experienced management, local operating costs were kept to a minimum. Rather, errors in business judgment and extravagant spending at the senior management level got Anson into financial difficulties. The experienced lumbermen who sold him the mills may have anticipated the depression and unloaded their properties before lumber prices dropped.

Anson had bought the mills, assuming a large mortgage on each, without waiting to consolidate the expense of the previous purchase before buying the next mill. Then he spent thousands of dollars building houses, churches, schools, mill enlargements, and improvements. His operations were spread out too far, and he had too many managers, drawing too high salaries — Hargreaves in Barrie; Brewster, Christie, Macaulay, Miller, and Sprague based in Orillia; and managers at each mill. Not only did these men earn large salaries, but the company provided them all with houses and unlimited expense accounts. On one occasion, for example,

Macaulay charged the company $100 for sodding a croquet lawn at his home in Orillia.

Anson did not draw a regular salary but took money out of the company in large amounts whenever he needed it — $5,000 here, $2,500 there. He also charged many personal expenses to the company: solid-gold scarf rings and cuff-links; money tied up needlessly and extravagantly in *Mittie Grew*; furniture purchased for Beechcroft; young deer shipped all the way from Byng Inlet for his deer park at Roche's Point. The labour bill for construction and landscaping at Beechcroft in the summer of 1872 amounted to $1,450, about two-thirds the monthly wages for ninety men working in the Waubaushene mill.

On the four large mills he bought, Anson was committed to making a payment in 1872 of $100,000 and annual payments thereafter of $40,000, at 7 per cent interest. The mills were not producing anywhere near enough surplus to make these payments, as is indicated by the statement of profit and loss on the mills of Georgian Bay Lumber (see appendix A). Although, as the statement shows, the company made a book profit of $20,391.86, the calculation of profit included $70,044.22 worth of lumber in inventory at three mills. Theodore Buck may have been happy with the figures from a manager's point of view, but Anson could not make mortgage payments with lumber piled in a mill yard, no matter how much it might be worth.

It is doubtful if Anson ever made any serious calculations of costs, sales volumes, or interest rates, before investing so rashly in the Georgian Bay mills; he definitely did not take into account the possibility of a decline in market demand for lumber. He was probably lured into buying the mills by the good luck he had with the Anson mill in Byng Inlet. When he began operating there, the US market for lumber was strong and prices were good.

No figures are available for the first year or two of operations at Byng Inlet, but a sales statement for 1871 reveals that the Anson mill shipped 5,949,118 board feet of lumber at prices ranging from $8 to $26.25 per thousand board feet, with an average of about $10.50. The profit on these sales was approximately $14,872 ($2.50 per thousand board feet).

On the basis of these profits, Anson reasoned naïvely that all he had to do was produce more lumber, through acquisition of more mills, and profits would increase proportionately. If he could produce, as he hoped, 80 million board feet at a profit of $2.50 per thousand board feet, he would net $200,000 annually. It was on the basis of such simplistic thinking that Anson's business decisions seem to have been made.

In 1872, it cost about $8 to produce one thousand board feet of lumber, taking into account timber dues and the cost of cutting and delivering logs to the rivers, river drives, towing booms, manufacturing the lumber, and loading it into ships. The domestic wholesale price of pine lumber was about $10.50 per thousand board feet, ranging from $6 for culls to $15 for prime-quality lumber. Lumber entering the United States was subject to a customs duty of $2 per thousand board feet, and freight costs ranged from $3.50 to $6 per thousand feet. Therefore only primegrade lumber could be sold profitably in the American market. Clear one-inch pine lumber retailed in Cleveland for $48 per thousand feet in 1872, but the wholesale price was only about two-thirds as much. Anson did manage to ship some reasonably good-quality lumber to Chambers and Co., getting $20 to $27 per thousand board feet for it, but he was able to ship only a few thousand feet at those high prices.

Transportation costs and customs duties were making it difficult for him to penetrate the American market. He did try to clear out the inventory of lower-grade lumber at his Canadian mills by shipping some to his own yard in Philadelphia, but it took a long time to sell the wood and the price received was very low. The cost of production remained constant throughout the 1870s, but price and demand fluctuated. When both demand and price began to drop sharply in 1872-73, Anson found himself in serious financial trouble. He had played Monopoly with real dollars and real property and was losing the game. But before he surrendered the dice, there was one more bit of "insanity" — to use his word — in which he felt compelled to participate. We examine this foolishness next.

Chapter Six

Politics Makes Strange Bedfellows

A.G.P. Dodge had not been long in Canada before he realized that business and politics were closely connected and that if he hoped to succeed in one he would have to participate in the other. Although the Canadian parliamentary system was foreign to him, although he knew practically nothing about the history of the country or the political philosophies of the two main parties, he did understand the importance of currying favour with those who held the reigns of political power, whoever they might be. And so he sought to ingratiate himself with politicians who sat on the government benches, in both the provincial and dominion parliaments. Because, when he came to Canada, Conservatives happened to be in office in both Ottawa and Toronto, Anson became a Conservative. But one has the feeling that he could just as easily have become a Liberal, if that party had been in power.

One of Anson's associates was the well-known Conservative Frederic William Cumberland, one-time engineer-architect, then managing director of the Northern Railway. Cumberland represented the large northern riding of Algoma in the Ontario legislature from 1867 to 1875 and in the Canadian House of Commons from 1871 to 1872, after which time dual representation was abolished. Anson's railway interests brought him into close contact with Cumberland, and they became good friends. Another associate was John Beverley Robinson, former mayor of Toronto and representative from Toronto in both the Ontario legislature and the House of Commons. Robinson was president of the Northern Railway. Another friend was Angus Morrison, Edinburgh-born lawyer after whom the Simcoe County village of Angus was named. From 1854 to 1863 Morrison sat as Conservative representative for Simcoe North in the legislative assembly of the United Province of Canada, and from 1867 to 1874 he represented Niagara in the Ontario legislature and in the House of Commons. He was a director of the Northern Railway and of Anson's favourite railway scheme — the Toronto, Simcoe and Muskoka Junction Railway. And finally, there was Anson's lawyer, D'Alton McCarthy. McCarthy would play an ambivalent role in Canadian politics, but in 1872 he was a strong supporter of Prime Minister John A. Macdonald and, indeed, was a Conservative candidate for Simcoe North in the Dominion election that year.

These men used their influence in government to help Anson secure the timber licences and mill site on the Magnetawan River, and they helped him obtain timber licences over and above those he bought in the timber sale of 1871. In gratitude Anson gave his political friends shares in his lumber companies. Robinson received five shares in the Georgian Bay Lumber Co. and was made a director, Morrison got ten shares in the Maganettewan Lumber Co., and McCarthy was given five shares in each of the companies Anson established.

If Anson needed the support of the politicians in his business and corporate dealings, they, in turn, were anxious to make use of him, not just for personal gain, but also for what Anson's investment capital could do for the Canadian economy and ultimately for Conservative political fortunes. Probably no individual had ever come to Canada and invested as much money in Canadian industry in such a short time as Dodge had. From Lake Simcoe to Byng Inlet, the woods were swarming with lumber-jacks, most of them in his employ. The economic spin-off from his lumbering operations was being felt everywhere. Consequently, by 1872, Anson was probably the best-known and best-liked resident of Ontario. He was charming and hospitable and spent money freely, qualities that endeared him to politicians.

It is not clear whose idea it was that Anson seek a seat in Parliament in the election of 1872. It may have been his own, for, as indicated earlier, he tended to be a mimic and may have wanted to join his friends in Parliament, the most prestigious men's club in the country. The Toronto *Globe* claimed that it was Cumberland and Robinson who persuaded him to run, but, regardless of who proposed his candidature, John A. Macdonald reluctantly supported it. He did not know Dodge, but because he had an innate suspicion of all "Yankees," a meeting with Dodge was arranged. The two men seemed to get along well enough. Afterward Anson wrote: "Since our first meeting I have felt that it would be a source of pride to belong to your clan and faithfully fight at your side."[1] Typically, Macdonald never revealed his true assessment of Anson.

York North was the only logical riding for Anson to contest. Beechcroft was in York North and Anson was well known and respected in the town of Newmarket. But winning a seat in North York would not be easy — virually impossible if he were to run as a Conservative candidate. The riding had been a Reformer-Liberal stronghold since the beginning of representative government in Ontario. The first secret meeting of the Reformer insurrectionists, who planned William Lyon Mackenzie's ill-fated march on Toronto, took place in Newmarket on 3 August 1837. After the brief rebellion, many of the rebels were incarcerated in the Old Scottish Kirk on Trinity Street in Newmarket. Prior to Confederation, York North was consistently represented in the legislative assemby by Reformers — Robert Baldwin, Henry Widdifield, and Thomas Mulock being the most notable. Since Confederation, Reformers had represented the riding in both the provincial and dominion parliaments.

But there were signs that the Liberals' grip on the riding was beginning to weaken. There was division between the pragmatic and doctrinaire wings of the party — "Grits" and "Clear Grits" — with evidence of much local jealousy and sectional rivalry. As a result, in a very close contest in the provincial election in March 1871, Newmarket lawyer Alfred Boultbee outpolled his Liberal rival, albeit by only five votes. At the instigation of George Brown, publisher of the Liberal organ, the Toronto *Globe*, the Liberal party organization had thrown over incumbent Dominion member James Wells, reeve of King township, and was about to replace him with James Parnham, a farmer from Holland Landing, as candidate in the next Dominion election.

In these circumstances, leading Conservatives thought the popular A.G.P. Dodge might stand a good chance of winning the seat. But Anson was an American citizen, resident in Canada for only three years, and therefore not eligible to become a British subject, a qualification necessary for sitting in Parliament. (There was no such thing as Canadian citizenship in 1872.) Macdonald removed this barrier to election by introducing a special piece of legislation[2] on 23 May 1872, naturalizing Dodge as a British subject. The only stipulation in the act, which received third reading and royal assent on 14 June — the day Parliament was dissolved — was a requirement that Anson take an oath of allegiance to the Queen within three months of passage of the act.

The citizenship matter was not the only obstacle to election Anson faced. Reformers in the riding might have been divided on philosophical grounds, and they might have disagreed among themselves on the choice of an appropriate candidate, but they were solidly united in their opposition to the Conservative government of John A. Macdonald. If Anson hoped to lure enough voters from the Reform cause to ensure his own election, he would have to distance himself from the Macdonald government. Therefore, it was decided that he would run as an Independent, and the Conservative party would not field an official candidate.

Anson had been working both sides of the political street since 1871. At the testimonial breakfast (referred to in chapter 3) held for him in Bracebridge in February 1871 he stated, though there was really no call for him to say so, that "he had not as yet allied himself to any of the political parties in this country" and that "he was carefully observing things and was resolved to identify himself with that party which was best calculated to promote the interests of the people."[3]

There may have been some sincerity in this statement, but, given Anson's close association with leading Conservatives in Toronto, one is led to believe that it was made either to prepare the ground for his candidature in York North or because he was being careful not to alienate the provincial Liberals, who were gaining strength and were on the verge of overthrowing the coalition government of the Conservative Sandfield Macdonald. Further, Muskoka was represented in Toronto by Liberal A.P. Cockburn; and H.H. Cook, with whom Anson, in concert, would later purchase timber licences in Muskoka, was also a Liberal. Cook was the provincial candidate for Simcoe North, where five of Anson's mills were located.

As early as April 1872, newspapers were beginning to mention Dodge's name as a possible candidate for York North in the Dominion election. Anson, who may have initiated the reports to test the water, said nothing to contradict them. Finally, on 19 June, the Toronto *Mail*,[4] a Conservative newspaper, announced that "Mr A.G.P. Dodge, the wealthy lumberer who has already invested millions in Canada, has finally yielded to the solicitations of leading men of all parties in North York to become a candidate in that riding."[5]

When the *Mail* made the announcement, it was by no means certain that Anson would run. He had been having second thoughts, mainly because he was experiencing financial difficulties, teetering, in fact, on the edge of bankruptcy; it had been because of his alleged wealth, which would be needed to wage a successful election campaign, that he had been encouraged to run in the first place. Moreover, he was having marital problems. Although he had spent a fortune building a Virginia-style mansion for his wife, Rebecca, at Roche's Point, she had refused to come to Canada and had gone to Europe for a year, taking their son with her.

Word that Dodge would not seek election reached Macdonald just four days after the dissolution of Parliament. The prime minister was perturbed. He wrote to Boultbee, Morrison, and Robinson, on 18 June, beseeching them to force Dodge to change his mind. To Boultbee, Macdonald wrote: "It is very vexatious to see Dodge behaving as he is doing . . . his friends ought to pitch into him strong."[6] To Angus Morrison he said: "I am told that Dodge is flattening out after all. This must not be allowed. Pray bring all the pressure that can be brought upon him. See John B. Robinson and others. He can resign at the end of the first session. After what he said and after getting his naturalization bill, if he backs down now he will destroy the confidence of everybody in his word and in fact will have wrecked his reputation for veracity."[7]

To John Beverley Robinson, Macdonald wrote: "You have great influence with Dodge. Pray don't let him back down from running in North York. This is all important."[8]

Anson's friends must, indeed, have "pitched into him," for on 21 June the Newmarket *Era* published the following letter from him.

> On my return from New York on Saturday last, I learned that there had been announced in your paper that I was not a candidate for the representation of North York. It is true that from the nature of my personal interest, I am not desirous of, at present, entering into political life; and feel, moreover, that I have too lately come among you to aspire to the honour of representing your riding. Yet finding so large a number of influential electors of all parties are desirous that I should represnt the constituency, I shall, if the elctors of North York

believe that in my position I can be a service to the riding, and the Dominion, most cheerfully offer myself as a candidate for their suffrages.[9]

By 5 July, Anson had formally declared as a candidate, and as a first step in the campaign he published, in all the county's newspapers, an address to the electors of York North. While he extolled the virtues of British institutions and pledged support for a few specific issues, particularly those identified with the goals of the Reform party — the Huron and Ontario Ship Canal,[10] immigration, and northern development — his main thrust was an attempt to dissociate himself from any party affiliation: "If returned by you to the legislature of this my adopted country, I shall go there untrammelled by a single pledge to any party," he told the electors. "I pledge you my word that I will support no measure, no matter by whom brought forward, unless, in my judgement, it is a good one . . . if the electors expect in me a strong, or mere party politician, they will be disappointed . . . I go with the earnest purpose of trying to do good, and will steadily seek to avoid being misled by mere party cries and issues."[11]

But the Toronto *Globe* and the Newmarket *Era* were not deceived by Anson's declaration of independence. With mixed metaphors, the *Era* proclaimed: "The savory nature of his candidature in Sir John A. Macdonald's organ at Toronto smacks so strongly of the government kitchen that we must confess it leaves little or no room in doubt the colours Mr. Dodge intends to sail under."[12] The *Globe* neither minced words nor mixed metaphors: "If he goes to Parliament now, he will get up and sit down as Mr. Cumberland tells him."[13]

Whatever flaws in character A.G.P. Dodge may have had, disloyalty to friends was not one of them. Having entered the race at the urging of his friends, he then devoted the same kind of zeal to winning the election as he had exerted in building his lumber business. He attended numerous public meetings and made speeches throughout the riding. He increased the tempo of construction and landscaping at Beechcroft, creating jobs for local residents to the degree that he was practically running his own "public works" program in the north end of the riding. He committed a great deal of money both for bribes to electors and for travelling expenses for those who lived some distance from polling stations. Farmer James Parnham could not possibly match the amount of money Anson was spending in the election campaign. Naturally he was accused of buying the election.

As nomination day — 29 July — approached, the campaign became bitter. Brown's *Globe* attacked Anson's character: "It is an insult to the electors of that fine Reform county on the part of the government to bring forth such a man," the *Globe* raged. "Mr. Dodge is wealthy and for the sake of writing M.P. after his name is probably willing to spend his money," the editor charged. "He evidently hopes with impunity to play the part of Tweed and Connolly in New York; but there is enough of sound principle in North York to send this Yankee Tory — the most contemptible kind of politician this world contains — back to Gotham, in spite of all the money he can spend."[14]

The editor of the Newmarket *Era,* who knew Anson and liked him was less vituperative. "I have reason to believe him a gentleman possessing good qualities and virtuous appearance," the editor observed, but nevertheless he advised Dodge "to reconsider the steps he [had] been advised to take."[15] Then, maybe in the future, "when he has [had] time and opportunity to become more familiar with the country and its institutions, the honest workers and liberal minded might endorse him and unite behind him."[16]

Faced with such criticism, Anson began to back away from his original declaration of "independence" and tried to convince the electors that, as a liberal-minded individual, he was really a Reformer. He rebuked the *Mail* for listing him as a government candidate. At least a dozen times in as many places in the riding, he avowed himself to be an out-and-out Reformer, and on several occasions he declared that his vote in Parliament would not be wanting to oust the corrupt Macdonald government from power. His objective in taking this line was, of course, to convince the Reformers in the riding, who were in the majority, that he could serve their interests in Ottawa better than James Parnham, the official Liberal candidate.

Interest aroused by the campaign attracted a larger attendance at the nomination meeting than on any similar occasion in the previous twenty years. Only two candidates were nominated: James Parnham, as the Reform Convention candidate, and A.G.P. Dodge, as an Independent. Anson's name was placed in nomination by a moderate Reformer, William Loucks from Port Hope, Ontario. Another Reformer, A. Ridell of Georgina township, seconded the nomination. Anson's strategy for attracting Reformers by claiming no party affiliation but holding Reform sympathies seemed to be working.

But a whisper campaign, conducted by his opponents, threatened to weaken his support among both Tories and Reformers. Some claimed that Dodge was divorced; otherwise why the big house at Roche's Point, but no wife? Still others said he was a drunk, a rumour that originated because Anson's campaign manager, Charles Ostrander, a clerk in the Orillia office, was an alcoholic, and Anson was frequently seen in his presence.

In an effort to sustain his credibility and to dispel some of the unsavoury allegations against his private and moral character, Anson wore a mantle of Christian piety during the election campaign. As it is difficult to assess the other features of this multifaceted man, so it is with his religious convictions. He was certainly a teetotaler and insisted on abstinence by his employees, Ostrander being the one exception. Not only were his mill managers and superintendents abstainers, but they were also extremely religious. His father, as indicated, was a professed and practising Christian. His older brother was a devout churchman, and a younger brother was an ordained Presbyterian minister. Yet one gets the impression that Anson's own godly demeanour revealed more style than real conviction. While Methodist and Presbyterian churches were appropriate for his mill workers and their families, such nonconformist institutions were not suitable for Anson. He recognized that power and prestige in Canada resided in the Church of England, and so it was to that denomination that he turned to demonstrate his own Christian fidelity. He subscribed ten dollars per year to the little Anglican church at Roche's Point, but it was principally in St Paul's Anglican Church in Newmarket that he worshipped on Sundays. So great was his sudden commitment to the Church of England's Thirty-nine Articles that he was invited to address the Anglican Synod, in Toronto, in the spring of 1872.

If his religious observance was mere play acting for political purposes, as many of his critics charged, Anson performed the role extremely well. He strictly observed the Sabbath, at least in Newmarket, even refusing to allow his groom to exercise his horse on Sunday. It became clear later that one of the reasons for his religious fervour was to win the approbation of Canon Ramsay, rector of St Paul's, and through the canon the support of his son, Dr Robert

Ramsay. Occasionally, after evensong, Anson would visit the rectory to discuss with the parson and his wife, over tea, such church-related subjects as Sunday schools — a favourite interest of Anson's father, but of no particular interest to himself, except for the purpose of impressing Ramsay.

The Reverend Mr Septimus B. Ramsay was born in London, England, in 1802. For many years, he was secretary in London of the Upper Canada Clergy Society, which later amalgamated with the Society for the Propagation of the Gospel in Foreign Parts. Through activities in these organizations, Ramsay developed a desire to perform missionary work in Canada; consequently he emigrated to Toronto in 1848. Bishop John Strachan appointed him rector of St Paul's in Newmarket, where he also had charge of several surrounding missions, at least two of which later became parishes. In 1867, Ramsay was made an honorary canon.

Canon Ramsay had considerable theological attainment to his credit, and he possessed what was described as "more than average pulpit powers."[17] One of his three sons, Robert, practised medicine in Orillia and was also proprietor and publisher of the Orillia *Northern Light*. The Ramsays were long-time Reformers. Dr Robert Ramsay had contested the riding of Simcoe East in the provincial election of 1871, but he was beaten by Conservative William D. Ardagh.

Septimus and Robert Ramsay had a taste for money and were attracted by the smell of wealth that emanated from A.G.P. Dodge. Their political convictions notwithstanding, the Ramsays jumped on Anson's bandwagon, even though they knew his political allegiance, and Anson, who must have sensed their real objective, but needing all the support he could muster, welcomed them aboard.

"I do wish that . . . you would caution Anson strongly against making acquaintance with boys who have no good principles, for I fear that he has friends . . . who will prove a great hindrance in his course of study,"[18] William Earl Dodge Jr wrote patronizingly to his father about his sixteen-year-old brother, Anson, in 1850. If, as the sages say, "the boy is father to the man," Anson's boyhood habit of attaching himself to unprincipled friends was carried into manhood, if his relationship with the Ramsays is any indication. For no pair of greater scoundrels ever wore a clergyman's collar or an editor's eye-shade than the Ramsays, father and son. Anson's boyhood friends may have ruined his education, but the unprincipled Ramsays would be responsible for destroying his reputation before his political career in Canada had ended. To what extent this character assassination was the product of his own doing, and to what degree the Ramsays were responsible, we may never know, but one fact is clear: a bankrupt Anson Dodge would leave Canada, but, because of his involvement with the Ramsays, he left with a shattered personal reputation as well.

As early as April, Robert Ramsay's *Northern Light* had been promoting Anson as a possible Reform candidate in the Dominion election expected in the summer. And, in early July, when other Reform newspapers were ridiculing Anson, the *Northern Light* supported him. It is not clear whether Robert Ramsay volunteered his services or Anson approached him because of his newspaper's backing, but Ramsay went to work on Anson's campaign in about mid-July. Ramsay's principal responsibility was to write newspaper articles in support of Anson and to speak at public meetings on his behalf. With their well-known Reform affiliation, the Ramsays provided Anson with the kind of endorsement he needed to nullify the savage attacks of the *Globe*.

Robert Ramsay worked for Anson for three weeks, during which time he stayed at the Royal Hotel in Newmarket, at Anson's expense. To reduce expenses, Anson proposed — according to Canon Ramsay — that Robert stay at the parsonage during the campaign, but, because Anglican clergymen were instructed to remain neutral in politics, the canon refused, lest it turn his house "into a sort of committee room."[19]

But the devious canon was not neutral. He worked almost as hard to ensure Anson's election as Robert did. Although, for the sake of appearances, he refused Anson's request to canvass in Newmarket, he did agree to communicate with "those at a distance."[20] He went every day to the committee room to count potential votes and to give advice on how "his own Grit party"[21] could be circumvented. He appeared with Anson before the judge who administered the oath of allegiance to the Queen.

Canon Ramsay acted as a messenger for Anson, delivering political messages from one end of the riding to the other. On several occasions he made trips to Orillia, on behalf of his son, who was also working for Conservative candidate D'Alton McCarthy, in a contest with Liberal H.H. Cook in Simcoe North. Canon Ramsay suggested that Anson invite the venerable John Cawthra, former resident of Newmarket and one-time Reform member in the legislative assembly of Upper Canada, to come to Newmarket to speak on Anson's behalf. The canon volunteered to go to Toronto, where Cawthra was then living, to seek his attendance. Ramsay found the old man ill and unable to travel, but Cawthra did send a message of support to Anson.

The most reprehensible activity in which the Ramsays engaged was conspiring with Anson to fake a character reference. Anson had been concerned for some time about what he called "the most damaging lies" about him being spread by overzealous Grits. He told the Ramsays that he had letters in his trunk from an old clergyman friend, the Reverend Dr Clark of New Jersey, and from other friends of his father, attesting to his good character. He wondered if somehow these letters might be used to counteract the attacks. He discussed the matter with the Ramsays, who, according to Anson, proposed that Dr Clark's character reference be modified by addressing it to Canon Ramsay in Newmarket and that the letter be published. But first, Clark's permission should be obtained. Anson later claimed that he telegraphed Clark and that his consent was given.

It is doubtful if Anson ever had a character reference from Clark or if a telegram was ever sent. Anson's father had a close friend, Dr R.W. Clark, who lived in Albany — not New Jersey. This Dr Clark delivered a fitting appreciation of the Hon. William Earl Dodge at his funeral in 1883. The initials of Anson's clergyman friend were S.A., not R.W. It was perhaps typical that in planning the deception, Anson remembered that such a respected family friend as Clark existed but forgot his first names — unless, of course, there were two Dr Clarks.

The letter purportedly written by Dr Clark was embarrassingly immodest and, consequently, all the more damaging to Anson, when it became known that he had written it himself. "Reverend and Dear Sir — Hearing that one of my dearest friends A.G.P. Dodge is a candidate for Parliamentary honours in your country, I cannot deny myself the pleasure of urging upon you the important claims he has upon your friendship and the entire confidence of your people," the mythical Clark began his letter to Ramsay. The note went on to describe in laudatory terms not A.G.P. Dodge but a saintly person who was an amalgam of Anson's father and his successful and highly respected brother William. "He is universally loved by all classes, especially the poor, as any man who has lived among us," Anson wrote about himself.

He was a princely giver to all good objects and aided the churches of all denominations with that broad liberality which so distinguishes him . . . Mr. Dodge [read brother William] was for years President of our Young Men's Christian Association . . . his name is in all the churches and his acts of kindness and philanthropy were extended to many places and people all over the United States . . . He is certainly worthily following the footsteps of his father who is the most prominent layman in the great Presbyterian Church of America . . . a man of more liberal, broad, Christian views cannot be found . . . an energetic, earnest life for the good of others and noble and high aims are his life's record. We can hardly say enough in his praise.

"Dr Clark" made a direct appeal to the electors: "You could not find a better candidate for bringing and pressing forward every good measure. I hope he will be valued as he should be by all his fellow countrymen."[22]

Anson gave his handwritten letter to a woman friend, who, according to the editor of the *Globe*, "corrected his bad English and bad spelling"[23] and copied the forged letter for the printer at the Newmarket *Courier*.[24] The letter, together with one supposedly addressed to him from Septimus Ramsay, acknowledging receipt of Dr Clark's letter and advising its publication, but actually forged by Robert Ramsay, was printed on a fly-sheet and distributed throughout the riding on the eve of the election.

The vote took place during the week of 9 August. Anson outpolled Parnham by 279 votes and could now add two more letters to the "alphabet." He was now A.G.P. Dodge MP. "The happiest day of my life. Dodge Forever,"[25] a jubilant Canon Ramsay telegraphed from Orillia, where he had discreetly gone the day the fly-sheet was distributed, so that he could later disclaim any participation in its preparation.

After the election, the Ramsays attempted to collect from Anson the money they thought their services were worth. According to Canon Ramsay, Ostrander had told them that if both Dodge and McCarthy were elected they should ask for $5,000, but since only Dodge won they should demand only $2,500 — $1,000 for Robert Ramsay's services and $1,500 in compensation for the business allegedly lost to the *Northern Light* because of Ramsay's involvement in the campaign. According to Anson, Ostrander, who died shortly after the election, told a different story. He claimed that the Ramsays had tried to make him sign a paper saying that he agreed on Dodge's behalf to pay the Ramsays $2,500. "I would not give them a cent. They are trying to blackmail you,"[26] Ostrander allegedly told Anson.

How much Anson actually paid the Ramsays is not known, but however much it was, they did not consider it enough. Robert Ramsay initiated a lawsuit against Anson, claiming a further $2,000, but the suit was subsequently dropped.

In the mean time, George Brown got wind of the sordid affair. He was incensed that the voters in York North had ignored his paper's warnings and rejected his candidate in favour of Dodge, whom he disliked intensely. Brown was especially angry because Anson's workers had distributed the influential fly-sheet at a public meeting which Brown addressed in Newmarket the day before the election. A vindictive person, Brown sought revenge. Doubtful of the authenticity of the letter from "Dr Clark," and suspicious about Canon Ramsay's involvement in the affair, Brown employed an investigator to ferret out the facts.

In a casual conversation with Canon Ramsay in Newmarket, the investigator, who was unknown to Ramsay, got him to admit that he "did not know a clergyman by that name [Clark] in the United States or elsewhere."[27] When the stranger walked away smiling, the canon suspected that he had been tricked: a few days later Ramsay received a message from Brown, informing him that he knew about the canon's complicity in the forgery and that he "must either clear [him]self or plead guilty."[28] Concerned now about his position in the church, Ramsay went to the *Courier* office, retrieved all the documents, including Anson's original, handwritten character reference, and delivered them to George Brown. Throwing himself on Brown's mercy, and hoping to exonerate himself, he told Brown all he knew about the matter. Since Brown now had all the evidence he needed against Dodge, he ignored the lesser culprits — Ramsay and his son — and waited for an opportunity to pillory Dodge.

Parliament opened on 5 March 1873. Before doing anything with his evidence, Brown waited to see Anson's true intentions: to support the Conservatives, as Brown suspected, or uphold his pledge to the electors, and vote with the Reformers. To be sure of Anson's vote, Brown sent a message to Anson through James D. Edgar, Reform member from Monck and a former legal editor for the *Globe*: "Either vote with the Opposition or Brown would release the evidence of forgery."[29]

On 13 March, Opposition Leader Alexander Mackenzie moved an amendment to kill a government motion to set aside the election in Peterborough West, won by a Reformer, allegedly through election irregularities. Anson ignored Brown's threat and voted with the government against Mackenzie's amendment. The next day Brown published the story about the forgery, framed in the context of one of the most rancorous denunciations of character ever made against a member of Parliament, even in an age when newspaper editors were not known for mincing words.

Brown ended his long attack on Dodge by mocking him with a parody of his own carefully crafted words of self-praise:

> Mr. Dodge may be a man of liberal views; but Christian liberality is hardly as broad as to give sanction to FORGERY. He may be a man of noble and high aims; but his attainments will hardly command reverence of mankind if they only extend to an adroit adaptation of other men's signatures. His name may be 'in all the churches'; but henceforth it will stink in the nostrils of all good and honest men. The man who resorted to forgery to gain his seat is not likely to care whether he redeems a single pledge or falsifies the profession that enabled him to cross the Parliamentary threshold.[30]

The unfortunate Anson was placed in the unenviable position of defending himself against allegations that were not only derisive but mainly true. He rose in Parliament on the Monday morning following the *Globe*'s exposé' and tried to defend himself by smearing Brown. In this, his first speech in Parliament, he violated parliamentary etiquette by quoting a personal conversation with another member, James Edgar. In an even worse violation, he accused Edgar of trying to influence his vote by intimidation. The Speaker was obliged to intervene.

Anson wrote a long letter to Brown, published in the *Mail*, that was almost as vindictive as Brown's article. To detract attention from his own guilt, he exposed the Ramsays, and he accused Brown of conspiring with them — "confederating with blackmailers and disreputable people"[31] just as he, himself, had done seven months earlier. He claimed that his lawyer had in his possession the original reference from Dr Clark, but the letter was never produced.

However, just three days after the *Globe*'s attack, a telegram from Dr Clark arrived mysteriously on the desk of the Speaker of the House of Commons, in which Clark — this time L.A. Clark — advised that he had received a copy of the *Globe* article. He deplored the defamation of Dodge's character, and he confirmed that Dodge had his permission to use the letter anyway he saw fit. No one seems to have pointed out that however fast mail service might have been in 1873, it could hardly have delivered a Friday edition of the *Globe* to New Jersey in time for Clark to send a telegram by Monday morning.

No matter what Anson said in self-defence, he could not deny having sketched his own portrait; and he looked foolish for having done so. For several weeks, the Dodge-Ramsay scandal was in every major newspaper in the province, support or condemnation of Dodge depending on the paper's political affiliation. Then the Canadian Pacific Railway scandal made the headlines and Dodge was forgotten, except in Newmarket, where the Anglican church's investigation into the conduct of Canon Ramsay kept the issue alive a little longer.

It was a sardonic application of "a Deo victoria" that Septimus Ramsay was the one punished most for the forgery affair. The day after publication of Brown's exposé', the Archbishop sent a deputation to Newmarket to look into the part played by Ramsay in the scandal and to determine if there were sufficient grounds to appoint a commission of inquiry. The committee, which met in the basement of St Paul's Church, questioned Ramsay and interviewed other witnesses. D'Alton McCarthy appeared on behalf of Anson. After carefully reviewing the evidence, the committee recommended a full investigation.

The Bishop of Toronto convened a commission in St James School, in Toronto, in early June. The body investigated seven charges against Ramsay. These included active participation in the election, participation in the forgery, lying about his involvement, and taking money from the plates of St Peter's Church, in Toronto (a charge, made originally in 1866, of which Ramsay had been cleared). Another charge — that Ramsay had been involved in an insurance fraud[32] was dropped.

The commission found sufficient evidence for further proceedings, to be conducted by a higher church tribunal. But the proceedings did not take place. The Bishop had had enough of Canon Ramsay. The congregation at St Paul's was demoralized and divided; the usually crowded church was practically deserted on Sundays. A petition, which asked the bishop to remove Ramsay, was circulated and signed by many leading members. Consequently, the bishop asked for the canon's resignation, which was submitted at the conclusion of the hearings in Toronto.

Canon Ramsay met with the wardens and members of the congregation to inform them of his action. Typically, his last words to his flock asked not for forgiveness but for money. He had lost all his property the year before, he was giving up an annual salary of $700 by his resignation, and he had incurred expenses during the repeated investigations. "I shall now leave you to your unfettered deliberations, which might be somewhat interfered with by my further presence,"[33] he observed as he left the church. Ramsay died two years later. In 1885, a forgiving congregation, with perhaps a touch of irony, dedicated a stained-glass window to his memory.

As for Anson's brief parliamentary career, it was not impressive. His first, unfortunate address to Parliament has already been discussed. What should have been his maiden speech, delivered on 20 March during the debate on the speech from the throne, was eloquent but was made in the midst of the forgery allegations and seemed to lack credibility. Anson spoke once briefly on an election bill, and Macdonald appointed him, because of his well-known advocacy of temperance, to a committee to examine liquor laws in Canada. His only contribution to his riding was to obtain the appointment in Schomberg of an agent for issuing marriage licences.

Anson spent little time in Parliament. The failure of his business in April, 1873 took him to New York, where he spent the next few weeks trying to salvage what he could of his dwindling assets. He wrote to the prime minister from New York on 24 July offering to resign his seat, but Macdonald, who needed all the votes he could muster to maintain the government in the face of the Canadian Pacific Railway scandal, asked Dodge to stay on.

The second session of Parliament convened on 23 October, 1873. Macdonald met heavy opposition because of his own involvement in the railway scandal. His majority of six gradually diminished, as one after another the Independents deserted him. By Monday 3 November, the possibility of Macdonald maintaining a majority in Parliament was slim. Then, in the middle of the afternoon's grim debate, A.G.P. Dodge, MP strolled into the House and, to the accompaniment of loud cheers from the government benches, took his seat. This was his first appearance of the session. He had returned out of loyalty to Macdonald and, perhaps too, because he was broke and needed the parliamentarian's salary of $1,000 per year.

Anson's vote could not save Macdonald's ministry. On Tuesday, two more supporters defected. By Wednesday, it was clear to Macdonald and the cabinet that they could not win a majority, if a vote were called. Consequently, the prime minister submitted his resignation to the governor-general in the morning and, at 3 p.m., when the House convened, announced his action. On Friday, Alexander Mackenzie formed a new government. With nothing to keep him in Canada, Anson left the country for good — as the Newmarket *Era* put it, "unwept, unhonoured and unsung."[34]

The editor of the Newmarket *Era* was one of the few Reform leaders in York North willing to give Anson the benefit of doubt as regards his political convictions; therefore he was more shocked than most by Anson's betrayal. To this editor goes the last word on Anson's brief business and political career in Canada: "Had Mr. Dodge been content to pursue the course which honesy and moderation suggested, his career might have been as brilliant as it has been eventful."[35]

Chapter Seven

Bankrupt

A.G.P. Dodge, like other members of his family, was typical of those nineteenth-century entrepreneurs who, as de Tocqueville and others pointed out, wanted to live well and loved business. The frontier and the Industrial Revolution had inspired ambitious dreams and provided unlimited opportunities for acquiring wealth. This spirit fired the ambitions of Anson's father and grandfather and allowed them to achieve unparalleled success in creating a commercial-industrial empire. But, whereas the older members of his family possessed the entrepreneurial skills to build their industries and the toughness to keep them in the competitive jungle of the nineteenth century, Anson lacked business cunning and perhaps even common sense. And, whereas his father and older brother, William,[1] practised caution and moderation in their pursuit of wealth, Anson was daring and recklessly speculative.

In addition to investing hundreds of thousands of dollars in mills and timber limits in Canada, Anson put thousands into other properties, in the United States. These purchases included a part interest in a lumber mill in Jacksonville, Florida; timber lands in Wisconsin and Michigan; property and timber lands in Georgia; and a planing mill in Batavia, New York. Nearly all the property was bought on credit, with the expectation that profits from the sale of lumber would pay for the investments. As Anson himself explained: "I predicated upon the then price of lumber in Canada large profits and went into very large purchases. I purchased perhaps a million dollars worth of real estate in Canada expecting a profit at the then price of 2, 3 and 4 dollars a thousand upon about a hundred million feet of lumber. If the market had remained as it was at that time I expected to pay for my real estate out of profits."[2]

But the lumber market did not remain where it was. Even while Anson was investing heavily, the market was showing signs of weakening. He did not make any profit at all in 1871, and after the worldwide financial crisis of 1872, he found himself in serious difficulties.

Anson's rather naïve investment approach, as outlined above, stands in marked contrast to his father's cautious approach to investing in an industry as volatile as the lumber trade. William Earl explained his unique method as follows: "For fear of errors, deduct one half; to be very careful, deduct one half again; in these days of speculation we will once more deduct one half; as everything appears to be going to the bottom, we will another time deduct a half; on account of the destruction of the United States Bank we will go another half. Half of this final amount will be the share of Phelps, Dodge and Co. Now deduct the original cost and the remainder will be profit."[3]

Had Anson done even one set of subtraction, he might not have sustained the great financial loss he would eventually suffer. Stringency in the money market in the United States and Canada affected Anson's lumber business in two ways. First, shortage of capital created a depression in the building trade, lessening the demand for lumber and lowering the price. Lumber that cost over $8 per thousand board feet to produce sold for less than $7 in the United States. Second, the shortage of money increased enormously the cost of banking accommodation. Accommodation paper, endorsed by other firms, was discounted by the banks at usurious rates of 15 to 18 per cent. Payment of such high rates of discount on a million-dollar investment absorbed all the capital Anson had and left nothing for the payment of loans and mortgages. By the spring of 1872 it was quite apparent that, if he were to stay in business, he would have to invoke the clemency of his creditors. By then his liabilities totalled $3,132,357.

Anson went first where he had always gone, when he needed financial assistance — to his father, and this time even to his mother. When he was young and just starting out in business, he frequently borrowed cash from his father, who treated Anson — and his other sons — no differently than any other creditors. Loans were recorded, interest was charged. The sons settled up with their father every first of January. This time, Anson's mother gave him some railway bonds which he sold for cash, and his wife advanced a few thousand dollars from her annual income. But the small amounts he received from these sources were not nearly enough to keep him afloat. So he began to sell property.

His father bought his interest in the Georgia Land and Lumber Co. He sold the half-interest in timber lands he owned in Pennsylvania to the firm of Dodge, James and Stokes for $125,000 in cash. He sold his lumber yard in Philadelphia, and he sold the planing mill in Batavia for about $40,000. In May 1872, he transferred to his father 9,000 of his 17,800 shares in the Georgian Bay Lumber Co., as security for advances his father had been making. But despite family assistance, Anson was still in deep financial trouble as the economic situation worsened.

It was at this time that Anson amalgamated his Byng Inlet mill with White and Co.'s Page mill to form the Maganettewan Lumber Co. of Ontario. White and Co. undertook to sell most of the lumber produced in Anson's other mills through its outlets in Albany, Oswego, and Chicago, and to guarantee him a profit of $2 per thousand board feet. With the decline in the price of lumber, this agreement was in jeopardy, and Anson faced the probability of bankruptcy.

Anson offered to sell White and Co. the Collingwood and Anson mills, but he wanted advances in cash, so that he could reduce his liabilities. White and Co. was not prepared to advance cash, but it assured Anson, time and time again, that because it held outside assets, more than enough to cover all its liabilities, its promissory notes would be secure. Samuel White Barnard, New York banker and a partner in White and Co., told Anson that he had large personal assets outside his interest in the company. From all the information Anson had at the time, he reckoned Barnard to be personally worth over $1 million.

Still Anson was not convinced that selling his mills to White and Co. for anything other than cash would be a good thing. He would be able to sell the promissory notes of White and Co. only at great discount. Credit was becoming more difficult to obtain. Other US lumber companies had already collapsed. It was unlikely, therefore, that he could obtain further advances on his already heavily mortgaged real estate, and without cash to pay operating expenses he could not continue to carry on his lumber business in Canada. But White and Co. needed Anson to stay in business, because, as commission lumber salesmen, it was dependent on the cheap lumber it was getting from his Canadian operations. If Dodge and Co. collapsed, White and Co. would soon follow.

The partners in White and Co. suggested that Anson go to his father for money, but he informed them that his father had already bought large amounts of property from him that he did not really want and would probably be indisposed to put up any more money. It was then that White and Co., which had been less than honest with Anson, came forth with a proposal to which the cautious William Earl reluctantly agreed.

The proposal was as follows. White and Co. would give promissory notes to Anson to the value of $400,000 — $200,000 for each of the Collingwood and Anson mills — extended over five years, with interest at 7 per cent. To secure payment of the purchase price and interest, William Earl would hold the land and premises in trust, but White and Co. would have the use and benefit of them until the notes were paid. On 12 October 1872, Anson transferred the properties to his father under the above conditions, and on the same day White and Co. executed an indenture guaranteeing its part of the agreement.

But this arrangement still did not solve Anson's immediate need for cash. Consequently another agreement was entered into by the three parties: William Earl would make certain advances in order to assist in the solvent liquidation of both A.G.P. Dodge and the firm of Dodge and Co. In return, White and Co. agreed to assign the land and premises of the two Byng Inlet mills and the Collingwood mill to William Earl. Anson agreed to transfer to his father all the purchase money and interest. Then Anson, with the financial backing of his father, agreed to advance White and Co. accommodation paper as it needed it. The purposes of the agreements were primarily to save Dodge and Co. from imminent collapse and, incidentally, to help White and Co. through the financial crisis, which it optimistically predicted would last about six months.

William Earl discussed the proposal with William Earl Jr, who thought it a foolish idea. But William Earl went along with it, because he wanted to help Anson. He did, however, insist on insertion of a clause in the second agreement that provided for the three mills to revert to him in the event that any one of the other two parties went under. A further condition, suggested probably by William, committed Anson not to increase his liabilities. The second agreement was signed on 17 October 1872.

Anson thought that the arrangement would carry him through the difficult period and that within months he would be back on his feet again. William Earl was not so sure. It was not until the second agreement was drafted that he discovered the enormous size of Anson's liabilities — $2,300,995 in the debts of Dodge and Co. and $880,420 in personal debts. Given the uncertain conditions in the money market and the steadily declining price of lumber, William Earl, who no doubt performed some of his famous subtraction, decided that he wanted more security against the great risk he was being asked to take.

A week and a half after signing the October agreement, William Earl made what was probably his first trip to inspect Anson's Canadian operations. By then his son was a member of Parliament (the election scandal had not yet erupted), workers were going into the woods for the winter's cut of logs, and the lumber companies were operating with the appearance of normality. Nevertheless, William Earl had D'Alton McCarthy draw up the necessary papers transferring into his name all the rest of Anson's Canadian holdings — his shares in the Parry Sound, Muskoka Mills, and Longford lumber companies, their associated timber limits, and other pieces of property.

By some oversight, the timber limits bought in conjunction with H.H. Cook at the public auction of 1871 were not included in the transfer. Anson later told the court that his father bought the properties from him, but the price paid was only the statutory $1 minimum needed to legalize the transfer. In any event, William Earl now became president of the Georgian Bay Lumber Co., and Anson, who still held 880 shares, was managing director. Anson believed the arrangement temporary.

Through the winter of 1872-73, Anson nurtured the false hope that all would be well. William Earl was faithfully keeping his part of the agreement by paying Anson's bank loans and mortgages as they came due, preventing the collapse of Dodge and Co. He had been accepting large amounts of paper advances from White and Co. for lumber, but he was confident that the paper was secure, because of the alleged personal wealth of the partners. In the paper exchange between the two companies, Anson's company held a balance in its favour of $348,285, all of it protected, Anson thought, by the sale of the mills and his lumber reserves. But it was all an illusion.

White and Co. had been using larger and larger amounts of borrowed money; it had not paid William Earl anything for the Byng Inlet and Collingwood mills, and the outside resources it claimed to possess when it signed the October agreement did not exist. When the shaky financial condition that began in 1872 culminated in the dramatic crash of 1873, White and Co. was wiped out. The lumber trade in both Canada and the United States was one of the first sectors to suffer from the crash. Canadian-based companies such as White and Co. and Dodge and Co., both heavily dependent on US markets, suffered more severely than some American companies, because of the duty of two dollars per thousand board feet that they had to pay on Canadian lumber entering the United States. In early May, Samuel White Barnard wired Anson in Ottawa, urgently summoning him to New York. Anson arrived to find that White and Co. would be forced into receivership. The paper of that firm that he held was worthless.

The dilemma now faced by William Earl was what to do about his son's company — to continue to sustain it or let it also fall to bankruptcy. With the collapse of White and Co., the October agreement was no longer valid; legally, he was no longer obligated to assist Anson. He had already advanced about $730,000 under the October agreement, primarily to pay bank loans in Canada and a loan of $100,000 Anson had obtained earlier from his cousin Isaac Phelps, under William Earl's guarantee. Georgian Bay Lumber and all the other major holdings in Canada, except Beechcroft and the Muskoka timber limits, were in William Earl's name. He would undoubtedly acquire outright ownership of the Collingwood and Byng Inlet mills as a result of White and Co.'s default on the October agreement and earlier indentures. The Tobyhanna timber lands in Pennsylvania were now owned by the family firm of Dodge, James and Stokes.

Although Anson still held personally, or in the name of Dodge and Co., a few pieces of US real estate, thanks to William Earl's shrewd foresight all the major assets were now secure in his own hands. Anson's debts were still large, but most of his creditors were American bankers. When bankers made loans or bought accommodation paper at discount, they had to assume risks. That was the way business was done. To continue financing his son would be tantamount to assisting these bankers. And so, confronted with a choice between family sentiment and business logic, William Earl chose logic. However soul-wrenching the consequences, it would be the ignominy of bankruptcy for Anson, a lesson from which William Earl probably hoped his prodigal son might benefit. On 23 May 1873, he cut off all further financial assistance to his son, and on the same day Anson informed his creditors that his company was insolvent.

The leading creditors met the following week to investigate the condition of Dodge and Co. and, perhaps not realizing that all the productive assets were now in William Earl's hands, came away convinced that there were "abundant assets to satisfy all liabilities, if time was allowed to work them up to the best advantage."[4]

In August, William Earl initiated proceedings against White and Co. to recover the Byng Inlet and Collingwood mills. In September, when it was clear that Anson would be declared bankrupt, he had Anson sign over to him the remaining 880 shares of Georgian Bay Lumber still in Anson's name. There is some question about the legality of this transfer, because, technically, everything owned by Anson as of 23 May should have been surrendered to the trustee. Fortunately, neither the trustee nor the creditors had the advantage of historic hindsight possessed by this researcher; they knew nothing about the transfer.

On 24 October 1873, the inevitable happened. Two banks — Consolidated National and National Broadway — filed petitions in the Southern District Court of New York, praying that A.G.P. Dodge, William J. Hunt, and Samuel Schofield (the partners in Dodge and Co.) be declared bankrupt. The declaration was made on 15 November. Then the legislation of 1867 — An Act to Establish a Uniform System of Bankruptcy throughout the United States — came into play.

The act required as a first step that the bankrupts submit an inventory of assets to the registrar in bankruptcy and that the creditors submit a statement of claims for verification by the registrar. Typically, Anson overvalued his estate, not deliberately, but because he had no realistic understanding of how much property was worth. For example, he valued his household furniture in New York at $10,000. When asked why he put such a high value on it, he explained: "I gave it no particular thought at the time. I put down $10,000 without any idea of estimating the different pieces of furniture. I said to myself, considering what I had spent that I must have $10,000 there at least."[5]

Ultimately, the trustee placed a value of $3,000 on the furniture, but it was eventually bought by William Earl at auction for only $1,943.75. Some of the overvaluation of the property resulted from the continuing depression which pushed prices of real estate well below what Anson had originally paid for his properties. Beechcroft, for instance, cost at least $60,000 but was sold in the depth of the depression for only $10,000. The low price of lumber deflated the value of the timber lands in Wisconsin and Michigan and the timber limits in Muskoka district. Consequently, assets on which Anson placed a valuation of $1,088,756, in the fall of 1873, were evaluated by the trustee a year later at $225,513.26. But it was Anson's estimate that the creditors had in hand when they met on 10 February 1874 to determine what was to be done about the bankruptcy.

The bankruptcy act required the creditors themselves to determine what was to be done with the assets, on a vote of the creditors representing at least three-fourths the value of the claims. Because William Earl, with proven claims amounting to $521,367.94, was by far the largest creditor, it was not difficult for him to get the committee to make the decision he wanted. Consequently, the committee resolved that the estates of A.G.P. Dodge, William J. Hunt, and Samuel Schofield be wound up and settled and that distribution be made among the creditors by a trustee under the inspection and direction of five creditors. The committee of five creditors was named at the meeting, and a trustee — John L. Caldwalder, a businessman in New York — was appointed.

On 14 February 1874, the court ordered Dodge, Hunt, and Schofield to convey all their properties to the trustee. On 29 July, the bankrupts, having complied with the court order, filed a petition with the court to have a full discharge of all their collective and individual debts, provable under the bankruptcy act. On 27 August, the first hearing into the petition for discharge was held in the office of the court registrar. The hearing was subsequently adjourned and reconvened ten times, the final testimony not being recorded until 10 April 1876. William J. Hunt, bookkeeper for Dodge and Co., and Samuel W. Barnard, banker for White and Co., each testified once, but the principal witness was Anson.

It was a humbling experience for Anson to submit, month after month, to searching cross-examination by court lawyers into his personal and business affairs. By and large, he answered the court's questions with candour or agreed to bring forward facts that were not available from memory or in records present in the court to the next meeting. Only once did he refuse to identify someone, even though his lawyer advised him that he had no choice.

On 19 May 1873, just four days before the declaration of insolvency, Anson drew a cheque for $10,758.67 on the American Exchange National Bank and deposited it in favour of the New York gold dealer Kennedy, Hutchins and Co. He then withdrew the gold and converted it into currency for, as he put it, "the liquidation of some borrowed money and in payment of personal loans."[6] He admitted under questioning that $1,200 had been used to pay his quarterly rent,[7] and several hundred dollars had been used to pay grocers' and tailors' bills and the wages of the gardener and servants at Beechcroft. But he refused to name the recipient of the remaining $8,000, until the court ordered him to do so. Then he revealed the whole unpleasant relationship, financial and personal, between his estranged wife, Rebecca, and himself since their marriage.

In the latter months of his lumbering operations in Canada, particularly when he was spending liberally for the election campaign in York North, there were times when Anson was broke. He borrowed sums of money, amounting to about $8,000, from Rebecca's account, without, as he put it, "any consideration paper given";[8] in other words, without Rebecca's knowledge. When it became obvious that he was heading into bankrutcy, he felt obligated to pay back the money. He placed the cash in an envelope and had it delivered to Rebecca. A true southern belle, with no experience in handling money, Rebecca did not know what to do with so much cash. She handed it over to Titus B. Meigs, Anson's former partner and now a parter in Dodge and Meigs and a trusted employee of the Dodge family. Meigs gave Rebecca a receipt for the money and advanced her sums from time to time on request.

In "laundering" the money, Anson unquestionably wanted to conceal the transaction from his other creditors, although technically it was legal, because the transaction had been made before his failure. It may have been commendable on his part to try to prevent his wife's name from being dragged into the public hearing. But there was more behind his mock-gallantry than that. He was trying, in fact, to shield himself (and perhaps Rebecca) from the wrath of Rebecca's stepfather by concealing the fact that he had used her inheritance for his own purposes. Rebecca had been left a considerable amount of property in Alexandria by her natural father. Prior to her marriage to Anson, the property had been administered by her stepfather, John B. Dangerfield. Dangerfield did not like Anson. He made it quite clear that he would not trust Anson with Rebecca's inheritance. To quote Anson: " Mr Dangerfield made a special point of settling the whole amount upon my wife, and in rather an unpleasant way, and I made a point to state definitely to him that under no circumstances would I touch a dollar of principal or interest; that I wanted it kept entirely separate from my affairs."[9]

Dangerfield put the property into a trust for Rebecca, with himself as principal trustee. Annual amounts were advanced to Rebecca, who turned the money over to Anson for management and safekeeping. Not surprisingly, from time to time, he used some of the income, estimated to be between $3,000 and $4,000, for his own purposes.

Despite his attempt to prevent his family's financial affairs from coming under public scrutiny, the court eventually uncovered all the embarrassing details, to an extent that probably would not have been necessary, had Anson simply paid his wife the $8,000 through normal channels.

The final testimony by Anson was given on 10 April 1876, but this did not conclude the case. With the public testimony now before them, several persons, primarily those who had purchased what turned out to be worthless Dodge and Co. accommodation paper, filed petitions with the court asking for a review of the earlier decision of the court registrar expunging their claims against the estate of Dodge and Co. And on 15 May the Farmers and Mechanics National Bank of Philadelphia, the second largest creditor, with claims amounting to $66,711.81, filed a complaint opposing the granting of a discharge of the debts. The bank registered seventeen specific allegations of violations of the bankruptcy act, but, in the main, it charged Anson with having given fraudulent preferences by making payments and assignments of property to certain preferred creditors — his father, his wife, and Willam Jay Hunt, who was given $1,000 in cash two days before the failure.

Hearings into these petitions dragged on through 1877, 1878, and 1879. Finally, on 16 June 1879, Farmers and Mechanics withdrew its complaint, but only after Anson, Hunt, and Schofield signed sworn statements denying any resort to fraudulence in the bankruptcy proceedings. On 3 July 1879, the trustee, with the advice and consent of the committee of five creditors, paid out a 4 per cent dividend to each creditor. But it was some months before what was left of Anson's one-time empire was finally distributed.

As for Beechcroft, Anson's magnificent country home on Lake Simcoe, the trustee held it in trust for three years. Anson's former gardener, William Lynn, was kept on as caretaker at a salary of thirty dollars per month. The trustee tried to rent the property, but without much success. T.H. Sheppard, older brother of W.J. Sheppard and a clerk in the Orillia office, had the good fortune to rent the house in the summer of 1874, for his honeymoon, for only five dollars a month. Finally, in July 1877, the house and property were sold to Walter Gillespie of Toronto for $10,000 in cash. Gillespie disposed of the property, one piece at a time, but kept the house and thirty-two acres for himself. In October 1885, he sold the house and lot to Toronto stockbroker Edmund B. Osler — later Sir Edmund — for $8,000. For the next twenty-five years, Beechcroft was Sir Edmund's summer residence. He left the property to his son Gordon, who, in turn, bequeathed it to his children, Pat Osler and Phyllis Aitken. Today, Beechcroft is still in possession of Sir Edmund's descendants.

Chapter Eight

Unhonoured and Unsung

With the signing over to his father of his shares in the Ontario lumber companies, Anson's connection with the Georgian Bay Lumber Co. came to an end. During the several years of the bankruptcy hearings, he lived on St Simon's Island in Georgia, where he managed his father's lumber interest. He was paid no regular salary. His father paid rent for a house on East 44th Street in New York in which Rebecca and Anson Jr lived, and he gave Anson enough money to, as Anson put it, "keep my family from starving in my absence."[1]

Anson was embittered by the bankruptcy experience. He believed that his father had broken the spirit if not the letter of the agreement he had made with him. He had transferred his Canadian properties to his father for a nominal sum, partly, he thought, as security for advances his father had been making to him, and partly to protect his Canadian companies in the event of the collapse of Dodge and Co. He expected to get his Canadian mills back, when the bankruptcy business had been settled, but William Earl did not return them. Nor did he allow Anson to return to Canada to manage the companies on the creation of which he had gambled and lost so much.

Anson expressed something of his resentment over the loss of his Canadian companies in a letter to Sir John A. Macdonald, written in April 1882 from St Paul, Minnesota: "Others are enjoying the fruits of my labours there [in Canada.]"[2] The "others" were his youngest brother, Arthur Murray, and his nephew, Cleveland H. Dodge; they were gradually acquiring control of Georgian Bay Lumber from William Earl. Years afterward, William Earl's grandson, Percy Dodge, would distort, through oversimplification, the complex arrangements discussed in the last chapter: "Anson allowed the Georgian Bay Lumber Co. to go to ruin. He must have lost a fortune through neglect. In desperation, Wm. E. took it back and gave it to his youngest son, my father [Arthur Murray]."[3]

Relations between Anson and other members of his family became strained after the bankruptcy. He came up to New York from Georgia to attend his parents' golden wedding anniversary celebrations in 1878, and he was recorded as being present at his father's funeral in February 1883. But he seems not to have attended any other major family function, such as a wedding or funeral, thereafter. He did, however, send a wedding present to Cleveland H. Dodge in October 1883. Sometime between 1883 and 1885, his long, unhappy marriage to Rebecca ended in divorce, understandably causing great discomfort (and further alienation) for a leading New York family like the Dodges, with their strongly held religious convictions.

Neither his mother nor his father trusted Anson after his Canadian misadventure. Both named their oldest sons as executors of their estates, with each succeeding son authorized to take over should anything happen to one of the others. Although Anson was the second oldest, neither parent named him as a first executor.

If William Earl was stern with his son, he was certainly not mean. In the letter to Macdonald cited above, Anson spoke about a "new fortune" he had acquired in the south. Since his father paid him no salary, he had no money to invest, and, therefore, the only way he could have acquired a fortune while in Georgia was by his father giving him one. One can surmise that William Earl eventually paid Anson more for the Canadian companies than the one dollar he originally paid at the time of transfer. The mills were worth about $1 million, but, counting the advances William Earl had made to Anson under the October agreement of 1872, they had cost him only about $520,000. He would have got some of this money back when the proceeds from Anson's other assets were eventually distributed by the trustee. Also, in his will, William Earl left Anson $250,000, the same amount he bequeathed to each of his other six sons.

Anson left his father's employ in Georgia in 1879 or 1880 and tried, apparently, to establish his own lumber business once more. His reason for writing Macdonald in 1882, after a silence of nine

134

years, was to solicit the prime minister's assistance in acquisition of a mill site at Kenora, Ontario, on behalf of his former and once-again partner, W.J. Macaulay, who had been vice-president of the Georgian Bay Lumber Co.

When Anson lost the company, Macaulay moved to Manitoba, and for a few years he operated a lumber mill at Rat Portage (Kenora). He sold this mill in 1880, and in 1881, in partnership with Anson, he planned to build another. According to what Macaulay told Macdonald, he and Anson had "bought a large quantity of timber lands in Minnesota tributary to Rainy River and Lake of the Woods."[4] In November 1882, they were reported to have had camps in the woods with gangs taking out logs. They hoped to take out that winter 18 million feet of logs, which they planned to float down Rainy River to Lake of the Woods to be sawn in a mill still to be erected "someplace on the line of the C.P.R."[5]

No record has been found of timber licences issued in Minnesota in the name of either Dodge or Macaulay. In the early 1880s, Canadian lumbermen are alleged to have engaged in wholesale illegal logging operations on government land in northern Minnesota; it may be that the Dodge/Macaulay operation, if in fact it did exist, was one of those operations. It appears unlikely that the lumber mill was ever built. No record of one in Macaulay's name can be traced.

Nor does it appear that Macdonald made any attempt to assist with acquisition of the lease. He did not answer Anson's letter, and a half-dozen attempts by Macaulay to meet with him in person failed. Macdonald's federal government held jurisdiction over the Kenora area only briefly, during a boundary dispute between Manitoba and Ontario. The issue was settled in 1883, after which time the prime minister would not have been able to assist with the lease anyway. In any event, it seems clear that another of Anson's ambitious lumbering schemes — perhaps the last — ended in failure.

In June 1886, at the age of fifty-two, Anson wed for a second time. His bride was thirty-year-old Rose Voorhees of Newell, Illinois. The couple was married in a quiet ceremony in the home of the bride's parents, with the Reverend Mr Wolsey Stryker, a Presbyterian minister and friend of Anson's, officiating. None of Anson's family was present. The couple took an extended wedding trip through Canada, but whether they visited the Georgian Bay area is not known. Afterward, they took up residence in Danville, Illinois.

Anson built a mansion on Vermilion Street in Danville, and he and Rose entertained lavishly. Unfortunately, during the first year of their marriage, the mansion was destroyed by fire, while a ball was in progress. The many guests escaped without having an opportunity even to save their wraps and coats. Anson rebuilt the house, but on a less costly and pretentious scale.

The second marriage was happy. By all reports, Rose was a talented and beautiful woman, a favourite of Danville society and very popular with all classes of people. She did not belong to any church, which may have been one of her special attractions to Anson, who was probably disenchanted with church people by then. (Dodge family records[6] show that William Earl Jr, a philanthropist like his father, made a couple of donations to a Presbyterian church in Danville in the 1890s; so presumably Anson maintained some connection with the church and kept in touch with his older brother.) Anson and Rose had one daughter, Julia, who later became Mrs Jay Garriott.

Anson did not do much during the thirty-two years he lived in Danville. He dabbled a little in real estate, but, by and large, he seems to have lived off his inheritance. He was frequently asked to run for public office, but, with the experience in York North perhaps uppermost in his mind, he always refused. During the Billy Sunday campaigns, Anson supported the "dry forces" and helped them banish saloons from Danville. But he lived quietly and happily with his wife and daughter in their elegant home, where they frequently entertained Rose's many friends, until she died in 1910.

For Anson, in later years, the myth of great wealth became his reality. Because he believed the myth, his acquaintances in Danville accepted it as reality, too. Sometime before his second marriage, he adopted the rank of general, a title by which he was known in Danville for the rest of his life. Contemporary newspaper biographies portrayed Gen. Dodge with glowing epithets: "a man of considerable wealth,"[7] "in his investments displayed keen discernment and unfaltering enterprise,"[8] "wisely used his time and talents for the benefit of his fellow men,"[9] "well developed intellectual powers,"[10] and "obtained considerable influence [in Canada] being elected a member of the Canadian parliament."[11]

The praise, as this story has demonstrated, did not compare favourably with the truth. There seems little doubt that the always-charming Anson was a popular figure in Danville. But, despite the accolades and generous words heaped upon him, not one journalist ever pointed to one significant achievement by Anson in all the years he lived in Danville. Biographical material about him frequently contains more factual information about his father and his accomplishments than about Anson. The most any journalist could write about Anson's efforts was that he was "a favorite in social circles where intelligent men gather for discussion of deep and vital questions,"[12] or "his advice and influence [are] often sought,"[13] or "he gives attention to a consideration of the momentous questions which are shaping the trend of American History."[14]

Over the years he spent all his inheritance. At the end, this man who sought to become one of the wealthiest lumbermen in North America left behind a last will and testament one hand-written page in length, containing 130 words. He bequeathed what little he had left to his beloved daughter, Julia — a few bonds and a *trust* fund of $50,000, established for him by his mother before her death in 1903. But for the percipience of his mother, Anson Greene Phelps Dodge would have died penniless.

The man who had been raised in a mansion on Madison Avenue and who had built two of his own idled away his twilight years in a small Chicago hotel called the Antler. In his final year, as a semi-invalid, he lived with his daughter in her modest home on Everett Street, where he died on 28 May 1918. His body was taken back to Danville by train the same night, and the next afternoon, following a simple funeral service, the remains were placed in the family mausoleum in Springhill Cemetery. Other than his daughter and her husband, no relative attended the funeral. According to a local newspaper, "there was no singing and no eulogy."[15] A.G.P. Dodge left the world much as he had departed Newmarket forty-five years earlier — perhaps not "unwept," but apparently "unhonoured and unsung."

Part Two:

Second Empire: The Dodge Family (1873-1896)

William Earl Dodge 1805-1883
– National Archives Washington 111-B-1781

Chapter Nine

Rebuilding: William Earl Dodge

Immediately after taking possession of Anson's Canadian companies, in 1873, William Earl began to reorganize them. White and
Co. was declared bankrupt by the District Court of New York on 5
November 1873. On 13 March 1874 it signed over all its Canadian
mills and timber limits to the trustees in bankruptcy. But William
Earl did not wait for the formal declaration of bankruptcy before
initiating proceedings to recover the Anson mill in Byng Inlet and
the Collingwood mill which he considered rightfully his because
of White and Co.'s violation of the 17 October agreement.

On 16 August 1873, Samuel White told Anson, in confidence,
that Clark and Co. of Albany had already commenced proceedings
in Canada against White and Co. to get judgment on some $100,000
due it in connection with the Canadian business. The object, according to White, was "to get judgment ahead of anyone else and
buy in the mills, lumber etc. in Canada."[1] Anson immediately told
his father about Clark's action. William Earl was alarmed: if Clark
and Co. succeeded with its claim, "they would cut out all others";
the mills in Byng Inlet and Belleville would be sold to pay the claim,
and "all the real value of the assets of White and Company would
be gone and the other creditors would get nothing."[2] William Earl
wrote to D'Alton McCarthy the same afternoon: "Pray, cannot this
be stopped and if not cannot you proceed to get judgment in my
favour as soon as possible?"[3] McCarthy followed the latter course.

On 23 August 1873, he filed, on behalf of William Earl, a bill
of complaint in the Court of Chancery in Barrie against both White
and Co. and A.G.P. Dodge to recover the $749,993.93 that William
Earl alleged was due to him for the balance of the purchase price

of the Anson and Collingwood mills and for advances he had made to Anson under the October agreement. On 3 March 1874, the court ordered White and Co. and A.G.P. Dodge to pay the claims.

Because, for obvious reasons, they defaulted in the payment, the court ordered the mills and timber limits sold at public auction. William Earl bought the Collingwood mill and timber limits for $90,000, and Titus B. Meigs, acting on William Earl's behalf, bought the Anson mill and timber limits for $11,000 at a sale held at the Barrie Hotel on 3 March 1875. Meigs held the mill in trust until a new Maganettewan Lumber Co. was incorporated, at which time he signed the mill and limits over to the company. Clark and Co. had been out manoeuvred.

The purchase of the mills by William Earl was, in a sense, a formality, because the purchase money was paid back to him as claimant against White and Co. The purchase was necessary, however, because sale by public auction was the only mechanism available to the court for settling the claim. Had William Earl not bought the mills, Clark and Co. or someone else could have acquired them at a bargain price.

The Page mill was a little more difficult to acquire, because it was the exclusive property of White and Co.; as it was not covered by the October agreement, William Earl had no lien on it. He did want the Page mill, however, because since 1871 it had been operated in conjunction with the Anson mill. Besides, he did not want a competitor to move into the inlet. To acquire the mill, he made a deal with the trustees, Alfred Wilkinson and George C. Peters, who were empowered to sell White and Co.'s assets to pay the creditors.

White and Co., which operated the Collingwood and the two Byng Inlet mills briefly (after the purchase from Anson) under the firm name of Page, Mixer and Co. had taken out two mortgages on the Page mill: one worth $75,252 with the Royal Canadian Bank, the other valued at $59,771 with the Bank of Toronto. When the trustees took over the mill, $93,317.11 was still owing on these mortgages. Under an agreement among the trustees, the banks, and William Earl, the latter paid the trustees $110,000 US for the mill, the trustees paid the banks, the banks discharged the mortgages, and William Earl received clear title to the mill and timber limits.

On 1 November 1876, the provincial secretary granted a charter of incorporation for a second Maganettewan Lumber Co. under authority of the Ontario Joint Stock Companies Letters Patent Act, 1874. The company, which now owned the Page and Anson mills along with the premises, wharfs, and timber rights, was capitalized at $300,000. Three thousand shares of stock, each valued at $100, were issued, 80 per cent of which were paid up in full by the transfer of the mill properties. William Earl, who held 2,920 shares was president, and Theodore Buck was secretary treasurer. William Earl brought two of his sons into the company — George Egleston and Norman White, each of whom were given twenty-five shares.

In March 1876, the Parry Sound Lumber Co. was reincorporated. The capital, the number of shares, and the share value remained the same as for the original company. The main difference was that William Earl replaced Anson as president and Theodore Buck became secretary treasurer. Because of the depressed state of the lumber trade, William Earl decided not to keep Parry Sound Lumber. He sold all his shares to J.C. Miller, MPP, in 1877.

The Muskoka mills had never been incorporated during the several years in which they had been operated by their various owners. The half-interest that Anson purchased from Lewis Hotchkiss in 1871 was sold to Anson's father in 1872. During the depression of 1873, J.C. Hughson also sold a portion of his half-interest: 30 per cent to A.H. Campbell Sr of Toronto, 20 per cent to J.S. Huntoon of Collins Bay, and 30 per cent to N.H. Salisbury of Albany. On 8 January 1875, the several partners received the first charter of incorporation. The share capital of the company was set at $300,000, divided into 3,000 shares worth $100 each. Eighty per cent of the capital was paid up by the transfer of the mills to the company.

William Earl was not interested in maintaining this property. Rather than keeping all 1,500 shares (a half-interest) for himself, he sold 720 to Meigs and 60 to Theodore Buck, keeping only 720 for himself. J.C. Hughson was president of the company, and J.S. Huntoon was superintendent.

In 1878, because of the continuing depression in the timber and sawn lumber trade, the company decided to reduce the share capital to $150,000. William Earl and Meigs disposed of all their shares to Campbell and Hughson. Hughson was the major shareholder until 1884, when Campbell bought his interest and became the sole owner.

At the time of his financial difficulties, Anson transferred his shares in Georgian Bay Lumber to his father, with the exception of five shares, which he kept. The other shareholders transferred their stock to William Earl, too, and all the managers and hangers-on left the company. McCarthy was kept on as company lawyer. After 1873, William Earl held 9,985 shares, Theodore Buck five, D'Alton McCarthy five, and Anson five. William Earl was president, McCarthy was secretary treasurer, and Buck was general manager. Buck and McCarthy were directors.

William Earl sold the small Laramy mill in Sturgeon Bay to A.R. Christie[4] for $20,000 in 1873. For an additional $5,864, he sold Christie 1,446 acres of timber land that Anson had bought in Tay and Tiny townships, and he deeded over another 479 acres to various parties for $6,343. The Longford mill was incorporated under a separate company (see below). Only the Port Severn and Waubaushene mills, the large timber limits in Haliburton and Muskoka held under licence to the Ontario government, and 16,554 acres of deeded lands in Muskoka and Simcoe counties were maintained under the control of Georgian Bay Lumber. The company kept *Thomas C. Street*, *Queen of the North*, *Dauntless*, and the two tugs, *Lilly Kerr* and *Prince Alfred*, but *Kenosha* sank in 1873 and *Mittie Grew* was transferred to the Parry Sound Lumber Co., reducing Georgian Bay Lumber's capital assets by another $17,000.

Because he suspected that Georgian Bay Lumber was overcapitalized, William Earl had Theodore Buck make a detailed inventory, with estimated values of all the property now owned by the company. Buck's inventory (see Appendix C) revealed a capital value of only $600,000. Consequently, in 1875, William Earl applied for reincorporation of the company under the Ontario Joint Stock Companies Letters Patent Act. A charter was granted in 1876, reducing the capital from the original $1 million to $600,000, divided into 6,000 shares of $100 each.

It seems that William Earl had difficulty in deciding which of his three younger sons — George, Norman, or Arthur — to make partners in Georgian Bay Lumber, the largest of the companies he took over from Anson. Initially, George and Norman were each given five shares, and 2,980 shares were registered in the name of their New Jersey company, Dodge and Meigs. But by 1877, it was clear that George and Norman would take over the Georgia and Pennsylvania lumber operations, so the youngest son, Arthur Murray, was brought in as a principal partner with his father in the Canadian companies.

The stock in the Maganettewan and the Georgian Bay lumber companies was redistributed. George and Norman were dropped; young Arthur received a large block: 1,500 shares of the Georgian Bay Lumber stock and 750 of Maganettewan. A year later, when the Collingwood Lumber Co. was formed, Arthur had a one-third interest in it. It is highly unlikely that William Earl simply gave the stock to Arthur, as there is no reason to suppose that he treated his youngest son any differently than the older ones. The stock was probably considered a loan,[5] which Arthur would have been expected to pay back from profits.

Because of the continuing low price for lumber, coupled with the large expenditures William Earl was making, Georgian Bay Lumber, as well as the other companies, produced very little, if any, profits throughout the 1870s. And because of the depression, the company's property had materially decreased in value. Consequently, William Earl decided to reduce the capital of the company a second time. In March 1879, he applied for supplementary letters patent for reduction of the capital from $600,000 to $400,000, divided into 4,000 shares of $100 each. The shareholders received two shares for every three shares previously held. William Earl then held 2,996 shares, Arthur 1,000, and Buck the remaining four. In 1881, William Earl transferred another 1,005 shares to Arthur, who, with 2,005 shares, was then the major shareholder in the company.

As mentioned earlier, William Earl bought the Collingwood mill at public auction for $90,000 in 1875, when the claim against White and Co. was being settled. But the original owners, Peckham and Hotchkiss, still held mortgages on the mill for $11,289 and $71,484 respectively. William Earl paid off the mortgages with cash and received clear title to the mill and timber lands in Simcoe County that were associated with it.

Before investing in the mill, William Earl instructed Buck to make a thorough investigation of the property. Buck's survey showed that the main mill was 88 feet by 145 feet. The mill had three steam-engines, one of 50 horse-power, the other two of 30 horse-power each. There were six boilers, each four feet in diameter and 27 feet long; an 86-foot-high brick smoke stack carried away the fumes from the boilers and burned waste. Eighty men worked in the main mill, which had one gang saw, one large and one small circular saw, a stock pony, an edger, and a trimmer. The mill cut about 8 million board feet of lumber per season, of which about 4 million feet was sold in the domestic market. The piling yard accommodated 7½ million board feet. Buck estimated that there was sufficient pine in the mill's timber limits to last five years, if the mill continued to cut at the rate of 8 million board feet per annum. In addition to the main mill, there was a timber mill for squaring timber. It had a 5½ foot circular saw and was capable of handling logs 65 feet in length. When in operation, the timber mill employed fourteen men. The main mill and the timber mill employed ten horses. There was also a planing mill, 60 feet by 80 feet in dimension. This mill had a double surfacer, a re-saw, and a shingle machine. It was powered by a ten-inch cylinder, 20-horsepower engine. When it was running at full capacity, twelve men and one team of horses were required. The property had a boardinghouse which accommodated twenty-four men, and there were several small houses. A large frame-house accommodated the mill manager, who, under White's brief ownership of the mill, was Harvey Mixer.

Having purchased the mill, William Earl was not sure what he should do with it, but he seems to have decided to follow Anson's original plan to incorporate the Collingwood mill into Georgian Bay Lumber; for the next three years the mill was actually owned by the company. Then an event took place in Georgia that induced him to incorporate the Collingwood property as a separate company. In 1877, the Georgia legislature passed a law requiring out-of-state corporations holding more than 5,000 acres of land to incorporate under the laws of the state of Georgia. This would affect William Earl's Georgia Land and Lumber Co., which owned 300,000 acres of yellow pine in Georgia and which had been incorporated in 1868 under New York law.

Two days before the law took effect, the Georgia Land and Lumber Co. turned over all its holdings to George Egleston Dodge, an unincorporated person and a citizen of New York. Litigation over title to the Dodge lands began in the Georgia state courts in 1877 and continued for many years.[6] During this time, it was by no means certain that the Dodges would be able to maintain their large holdings in Georgia. So William Earl, always thinking about the welfare of his sons, decided that the Waubaushene and Collingwood mills should form separate companies in case the Georgia property was lost, necessitating the distribution of the Canadian properties among two or even three of his sons.

In October 1878, William Earl bought the Collingwood mill and timber lands back from Georgian Bay Lumber for $150,000, and he held them in trust pending the formation of the Collingwood Lumber Co. Letters patent were issued in December, granting a charter for a company with share capital of $150,000, divided into 1,500 shares worth $100 each. William Earl kept 1,090 shares, Arthur Murray received 375, and — maintaining the practice of profit-sharing started by Anson — William Earl divided the remaining thirty-five shares among the senior officers. The stock was fully paid up by the transfer of the property to the company. David Cooper was moved from the Waubaushene mill to manage the company.

As with the other mill properties that William Earl acquired from Anson, the Longford mills were in a financial and corporate mess. Anson held them in partnership with John Thompson; there was no incorporated company. The mills, being on the line of the Toronto, Simcoe and Muskoka Junction Railway and having easy access to timber in Haliburton and Victoria counties, were a potentially profitable asset. With Anson's bankruptcy imminent, William Earl could see no reason why this valuable property should fall into the hands of the trustee. Consequently, he sent Titus B. Meigs up to Ontario to arrange for purchase of Anson's half-interest in the mills. The indenture confirming the sale was signed in Barrie on 8 May 1873, just two weeks before the failure of Anson's company.

The indenture was drawn up hastily, and, typically, Anson did not know exactly what the firm's liabilities were. Meigs did learn that Thompson and Co. (the firm name under which the mills operated) had incurred liabilities of at least $110,000, principally on a loan from the Dominion Bank. Anson also had personal indebtedness of $14,000 to a man named Melville Miller for a portion of the purchase money for the property.

Meigs agreed, on William Earl's behalf, to buy Anson's half-interest in the firm for $75,000 and to assume Anson's share of the firm's liabilities and his personal indebtedness to Miller. By the terms of purchase, Meigs paid Anson $5,000 in cash and agreed to pay the balance in half-yearly instalments of $10,000. Because Anson had no formal statement of the liabilities, and to protect William Earl, Meigs insisted on inclusion of a clause that provided for a reduction from the purchase price of one-half of the value of any indebtedness that exceeded the alleged $110,000. William

Earl's distrust of his son's accounts was justified, because when he later had an opportunity to examine the books of Georgian Bay Lumber he discovered a charge of $24,000 against Thompson and Company for freight on lumber and other services rendered.

On 17 March 1876, a charter for the Longford Lumber Co. was granted. The share capital was $300,000, divided into 300 shares of $1,000 each. William Earl held 140 shares, John Thompson 148. The remaining twelve shares were divided among Buck, Meigs, and Alexander Thorburn, the company's bookkeeper. The stock was paid up in full by the transfer of the mills and timber limits. William Earl and Meigs were directors. Thompson was president, even though it was William Earl who gave his personal guarantee to the Dominion Bank for the company's indebtedness.

In 1878, at about the same time as he sold his interests in the Parry Sound and Muskoka Mills companies, William Earl decided to withdraw from the Longford Lumber Co. as well. But the firm owed him a personal debt of $54,000, it owed Georgian Bay Lumber $24,000, and it now owed the Dominion Bank $120,000, which William Earl had been guaranteeing.

To extricate himself from the company, William Earl made the following arrangement with John Thompson: he (William Earl) would accept $25,000 in full satisfaction of the $54,000 owed to him personally; he would pay Georgian Bay Lumber $20,000 in settlement of the $24,000 debt; and he would pay $100,000 to the Dominion Bank. In return, he received a mortgage for $145,000 on the mills and timber limits. Thompson agreed to pay the $45,000 portion of the mortgage in two instalments of $22,500, plus interest, on 1 January of both 1884 and 1885.[7] The $100,000 balance of the mortgage was to be paid in four yearly instalments of $25,000, beginning on 1 January 1880. Thompson then became the sole owner of Longford Lumber.

Through an oversight, Anson's share of the Muskoka timber limits bought in conjunction with H.H. Cook in 1871 had not been signed over to William Earl before the bankruptcy. Subsequently, the trustee sold them at public auction for much less than they had originally cost. Cook bought the limits and sold them to William Earl along with several square miles of additional limits that Cook owned outright in Muskoka and Haliburton. William Earl turned some of the limits over to Georgian Bay Lumber, some to Muskoka Mills Lumber, and the remainder to Longford Lumber.

While William Earl was reorganizing the corporate structures of the companies and putting them on sound financial footings, the mills continued to operate under the capable direction of the always reliable Theodore Buck. The head office was moved from Barrie to Waubaushene in 1873. Buck built a large, permanent home for himself on the hill behind the mill, overlooking Matchedash Bay. A two-storey office building was constructed. With William Earl's concurrence, Buck established a new adminstrative structure for the reincorporated companies, Anson's pals having by then departed. Some of the officers and senior employees engaged by Anson were kept on, but Buck brought in or promoted some new men.

Legally each company operated independently, but in practical terms, they were all managed from Waubaushene, Buck being general manager and secretary treasurer. J.H. Buck managed the Byng Inlet mills, Andrew McNeilly was superintendent of the Port Severn mill, David Cooper managed Collingwood, and J.C. Else was superintendent of the main mill at Waubaushene. W.H.F. Russell took over from his father as manager of the main store at Waubaushene in 1876; W.J. Sheppard managed the Port Severn store. Henry L. Lovering replaced Levi Miller as manager of the woods operations.

Lovering had come to Medonte township from England with his parents in 1842, when he was just seven years old. In October 1850, he commenced lumbering with James Sanson at Port Severn, earning a salary of five dollars per month. In 1852, he moved to the head of Lake Superior, where he was engaged in logging operations in the vicinity of what is now Duluth, Minnesota. In 1857, he returned to Ontario and became associated with A.R. Christie at Port Severn. When the Port Severn mill was sold to A.G.P. Dodge in 1871, Lovering joined Georgian Bay Lumber, and he remained with it for forty years. He also operated his own shingle mill in Coldwater.

In 1878, Buck relinquished his responsibilities as secretary treasurer to James Scott, a bookkeeper who had been employed by Anson Dodge since 1872. Scott was born in Tyrone County, Ireland, in 1839. At the age of six, he emigrated with his parents to Kingston, Ontario, where he attended elementary and secondary school. After teaching public school for a time, Scott joined the Royal Canadian Bank. When that bank collapsed in 1869, he took a position with the Crown Lands Department in Toronto. Three years later, because

W.H.F. Russell
1857-1933
– courtesy
William Russell

James Scott
1830-1902
– Huronia Museum

he knew the workings of the department so well, Anson hired him as senior bookkeeper in the head office in Barrie. He moved to Waubaushene when the headquaters moved in 1873. A very competent secretary treasurer, Scott performed much of the routine legal work that had formerly been done by D'Alton McCarthy. Later Scott was made vice-president.

All the men employed at the managerial level were devout Protestants, competent and dedicated workers, loyal to William Earl, and, without exception, temperance men. In fact, as in Anson's day, it was a stipulation with all employees that they abstain from intoxicating drink while in the company's employ. Lovering and Scott were the only Canadians employed at the senior management level. The senior officers were given a few shares in the companies, which provided them with incentive bonuses — when there were profits — and allowed them to sit on the boards of directors of the several companies, so that quorums for board meetings were possible without William Earl's presence in Canada.

Waubaushene, being the headquarters of the Dodge companies, was the largest of the mill villages. It was also the most attractive. There were two boarding-houses accommodating sixty men, and twenty-three houses accommodating thirty-nine families.(Port Severn had three boarding houses accommodating eighty men, and twenty-six houses accommodating thirty-six families.) Visitors and tourists who arrived in Waubaushene on the Midland Railway discovered a charming and neat little village of about five hundred inhabitants, the look of order being entirely due to the taste and energy of Theodore Buck.

In a eulogy to Buck, delivered in 1881, William Earl said: "He was a man of marked business habits, securing in all ways constant order and system in every department so that the villages under his control were models of sobriety and order. Strict rules of temperance were always maintained and no one allowed to sell in any place where he had oversight."[8] The grounds around Buck's own hillside home were beautifully landscaped, laid out with willow trees, flower beds, and shrubs. This set the standard for the residents in the smaller houses. Each residence was surrounded by a white-washed picket fence, all the fences of uniform height. Behind the fences lay trim lawns and attractive flower gardens. The windows of the houses were covered with netting in the summer to protect the inhabitants from the droves of mosquitoes that hovered around

Matchedash Bay. The residents were rewarded for their efforts by giving Waubaushene a reputation for its "thrift, beauty and lumber business."[9] William Earl, through Theodore Buck, made two significant management changes after Anson's departure — one practical, the other symbolic. First, the practice of operating the mills on Sunday was immediately discontinued. Second, because Anson's political behavior and business inefficiency had made a travesty of "a Deo victoria," the elaborate but affecting seal and motto of Georgian Bay Lumber were dropped. The new seal, depicting a fallen pine tree beside its chip-surrounded stump, conveyed unequivocally the purpose of the company. There was no need for a motto.

Reflecting Buck's own personality and style, the nineteenth-century Protestant ethic permeated Waubaushene, making it a sober and industrious place. During the milling season, the waterfront was a hive of activity from 6 a.m. to 6 p.m., six days a week. After 1875, when the Midland Railway reached the village, long trains of cars were loaded daily in the lumber yard, while schooners and barges were being loaded at the wharfs. The Midland Railway had six tracks running through the lumber yard and a loading track extending along the wharf, where barges unloaded lumber brought from Port Severn, four miles away. In October 1880, one visitor counted 106 railway cars in the yard being loaded with lumber and square timber destined to eastern markets.[10] So systematic was Buck's organization that he knew the location and quality of every pile of lumber in the yard, so that when a sale was made he could fill the order without leaving his office.

The steamer *Magnetawan* called at Waubaushene weekly on its regular scheduled trip around Georgian Bay, while the company's own schooner, *Thomas C. Street*, brought hay, oats, flour, salt pork, and other items of food and equipment for the village and logging camps. This material was stored in two large warehouses erected at the water's edge, at the bottom of what is now William Street. To accommodate the many visitors who went to Waubaushene either on business or to sightsee, the company erected the twenty-six-room Dunkin House. Naturally, it was a temperance hotel, getting its name from Christopher Dunkin, the Quebec member of Parliament who wrote the Canada Temperance Act of 1864, commonly known as the Dunkin Act. Guests paid one dollar per day for room and board.

Waubaushene c1881   – Simcoe County Archives

Willow Street, Waubaushene 1914   – courtesy Margaret Bell

Church services and Sunday school were held every Sunday — the three Protestant denominations sharing the company-built church. Roman Catholics continued to worship in the small stucture erected during Hall's ownership of the village.

Because of the increase in population, a new enlarged school house, twenty-four feet by forty feet, was built in 1880. It had only one room, but adjustable desks could accommodate the varying ages and sizes of the pupils.

On the few occasions when William Earl visited Waubaushene, he stayed with Buck. After 1878, Arthur Murray began to take an active interest in the management of the companies and generally spent some time at Waubaushene in the summer. But, by and large, management and operation of the business were left entirely in the hands of Buck, in whose judgment William Earl had complete confidence.

The depressed state of the lumber and timber trade that started in 1873 continued throughout most of the decade; consequently, William Earl's Canadian companies were largely unprofitable. But having been engaged in the lumber business for upward of forty years, he knew that lows in the trade were only temporary and that sooner or later both the demand for lumber and the price would improve. As mentioned earlier, he sold his interests in the Parry Sound, Muskoka Mills, and Longford lumber companies because of the depression, but he kept the other five mills and operated them, even at a loss, against the day when the trade would pick up.

Several factors contributed to stagnation in the Canadian lumber trade through the 1870s. Poor economic conditions coupled with unsettled public matters in the United States kept the price of lumber in American markets unremunerative. Consequently, there were few shipments of lumber from Canada for cash, and lumber shipped for sale by commission was either held over by consignees until the price improved or sold without profit.

Canadian producers had to overcome other obstacles before they could compete in the already weak American market. First, there was the customs duty on all Canadian lumber entering the United States. Second, low freight rates by water and rail from Michigan enabled producers in that state, at comparatively small cost, to supply leading centres of distribution, thereby overstocking the northern market. Third, and even more harmful to Canadian producers, unprecedented low railway freight rates throughout the

United States allowed Michigan producers to deliver lumber to points throughout the country that had formerly been supplied from northern centres of distribution, further lessening the demand at the usual Canadian markets of supply.

The distribution problem did not concern William Earl as much as it did some other Canadian producers, because he had access to American markets for his Georgian Bay lumber through Dodge, Meigs and Co. and Dodge, James and Stokes, but freight rates and the tariff were onerous burdens on his Canadian operations. Never one to wait for events to unfold, he attacked these two encumbrances with characteristic drive and cunning.

He first solved the problem of freight rates. Until the railway reached Waubaushene, the principal means of transporting lumber to American markets was by barge and schooner. Anson had no systematic way of shipping lumber, relying on a host of carriers, some of which charged exorbitant rates. William Earl reasoned that it would be cheaper to contract with one carrier to handle all the lumber for American markets. Accordingly, he renegotiated the agreement Anson had with Capt. Isaac May, owner of the four barges referred to in chapter 5.

For the shipping seasons of 1874 and 1875, Georgian Bay Lumber guaranteed May delivery on 40 million board feet of lumber each year from the Waubaushene, Port Severn, Muskoka Mills, Parry Sound, and Byng Inlet mills. The rates to Buffalo from Waubaushene, Port Severn, and Muskoka Mills were $3.75 per thousand board feet and from Parry Sound and Byng Inlet $3.50. The rates to Cleveland from the same mills were $3.25 and $3 per thousand feet, respectively. The company guaranteed full barges on each trip and agreed to haul the barges from the mill loading docks to open water at no cost to May. These rates went a long way toward offsetting the low rates paid by Michigan producers. After the coming of the railway, the company sold *Queen of the North* and *Dauntless*, but, as indicated, it kept *Thomas C. Street*, primarily for its own use. The contract with May was discontinued.

The other problem — the American tariff — was more difficult to solve, but William Earl approached it with characteristic determination. Because the Reciprocity Treaty of 1854 (to 1865) had stimulated the lumber industry in Canada more than any other measure, William Earl concluded that renewal of the treaty would once again lift the Canadian trade out of the doldrums. To this end, he used his influence in the US Congress and his network of international connections to try to change government policy in both the United States and Canada. In 1873, he encouraged the Liberal government of Alexander Mackenzie to try to negotiate another reciprocity treaty, but, unfortunately, the victory of the protectionist northern states in the Civil War had upset the regional balance in American politics, making the treaty impossible to achieve.

But the world-wide depression in the latter half of the 1870s brought about a shift in US attitudes toward free trade. Sensing the change and hoping to accelerate it, William Earl hired two of the most able writers in the country to make, as he put it, "a strong impression on the public mind,"[11] through newspaper and magazine articles promoting free trade with Canada. In September 1877, he passed this intelligence on to George Brown, publisher of the Toronto *Globe*, who had conducted the earlier unsuccessful free trade negotiations with the US secretary of state — the same George Brown who had engineered Anson's character assassination in 1873.

If William Earl knew about Brown's attack on his son, it did not stop him from writing a confidential and friendly letter to the publisher, encouraging him to try for another reciprocity treaty. "If you could get some other strong man to act with you I am very confident that the session [of Congress] will not pass without success."[12] But before the negotiations could begin, John A. Macdonald replaced Mackenzie as prime minister after the election of 1878, and he subsequently introduced the National Policy of protective tariffs.

Had William Earl's efforts been successful, he stood to benefit substantially; nevertheless there seems to have been a quid pro quo for assisting the Canadian government in its free trade bid. In 1877, the Mackenzie government issued a timber licence, covering many square miles, to Georgian Bay Lumber for three Indian reserves — French River, Magnetewan, and Naiscontiong — for a total bonus of only $500. Lumbermen paid an average bonus of $200 per square mile at the provincial timber auction held the same year.

By 1880, the American lumber market began to exhibit something of its former activity, with considerable advance in prices, prompting Ontario's commissioner of crown lands to report to the legislature: "The trade has at last emerged from the gloom which has so long overshadowed it."[13] He predicted that for years to come transactions would be fairly remunerative to shippers and dealers generally. At the same time, the British market for square and waney timber improved, yielding the highest prices ever for good-quality timber. Regrettably, in 1881, just when the fortunes of Georgian Bay Lumber began to improve, two disasters struck: on 12 May Theodore Buck died, and on 15 August the Waubaushene mill burned to the ground.

Buck's death came as no surprise. He had been suffering from some debilitating disease — probably cancer — for two years. William Earl delivered the eulogy at the funeral, in Mountoursville, Pennsylvania. As a more lasting tribute and in recognition of Buck's deep religious conviction, William Earl and Arthur Murray Dodge erected a church in Waubaushene dedicated to Buck's memory. This simple, but beautiful, little frame church, lined throughout with local ash, is still in perfect condition, providing a place of worship for the major Protestant denominations (now Anglican and United), as was intended when it was built in the summer of 1881.

Buck's death forced reorganization of the administration. The man employed to replace Buck was Jesse Peckham of Newmarket. Peckham was born in New York state in 1838, one of a family of eight children of Quakers Stephen and Mary Peckham. In 1854, Jesse came to Canada with his parents and settled at Warrington, near Stayner, where Stephen Peckham built a sawmill on the line of the Northern Railway. In 1862, the family moved to Newmarket, probably because of the active Quaker community there. From Newmarket, Stephen Peckham supervised his lumber business, which included mills in both Canada and the United States.

Jesse and his two brothers-in-law, Isaac Hoag and Edwin Stocking, were associated with Stephen in the lumber business, their largest mill operating at Collingwood in partnership with Lewis Hotchkiss. As indicated earlier, the Collingwood mill was sold to Anson Dodge in 1871; Jesse subsequently formed a partnership with Isaac Hoag — Peckham and Hoag Lumber Merchants. When this partnership dissolved, Jesse left Newmarket for a number of years. He returned in 1878 and bought shares in the Newmarket Tannery, which his father had established in partnership with a man named Parks in 1877. Jesse became business manager. He seems to have been a restless individual, not sticking with any line of employment for long. In 1881, hoping to re-enter the lumber business, Jesse was negotiating with Georgian Bay Lumber for a management position when Buck died, and, to his surprise, William Earl offered him the general managership. He moved to Waubaushene in June.

The Union Church, Waubaushene

Jesse Peckham was well prepared for his new job: he had considerable knowledge of all phases of the lumber business, having been practically born into the industry, and he had some management experience. He was a fine-looking man, unmarried, and had stirling qualities of character. Moreover he was a member of the Presbyterian church in Newmarket, having left the Society of Friends at a young age. He was also a teetotaller. But he proved not to be as efficient a manager as the man he replaced and remained with Georgian Bay Lumber for only five years. Meanwhile he and James Scott were each given five shares in the company, and both served as directors.

William Earl made other changes in the company in 1881. He transferred another 1,000 shares of stock to Arthur Murray, giving him controlling interest, with 2,005 shares, signalling the day when Arthur would take over full responsibility. Also, in 1881, William Earl brought his grandson, Cleveland H. Dodge, into the company. Born in 1860, Cleveland was the second son of William Earl Dodge Jr. After attending boarding school in Easthampton, Massachusetts, Cleveland went to Princeton University, where he was a classmate and later a close friend of Woodrow Wilson. Cleveland and his older brother, William Earl 111, graduated from Princeton in 1879. Following the principle of primogeniture, William Earl 111 joined Phelps Dodge and Co., while Cleveland, like his uncles before him, went to work in Dodge, Meigs and Co. to learn the lumber trade. When he turned twenty-one, his grandfather gave him five shares in each of the three Canadian lumber companies.

If Buck's death was expected, the fire that destroyed the Waubaushene mill was not. At four o'clock on Sunday afternoon, 15 August, someone discovered a fire in the mill, but before anything could be done about it, the building was a mass of flames and totally consumed. The cause of the fire was never determined, but since the mill was not in operation it was assumed that whatever ignited the blaze had been smouldering since Saturday afternoon. The building was fully insured. Construction of a new mill began almost at once.

The new mill, which cost an estimated $85,000, was ready for the spring cut. The main building was 126 feet by 70 feet — considerably larger than the original building; the lath mill measured 20 feet by 60 feet, and the filing room was 16 feet by 30 feet. An attached engine room, which housed two 250-horse-power steam-engines, was 20 feet by 32 feet. A boiler room, separated from the main building by a brick wall, was 30 feet by 41 feet; it contained six boilers installed by Thomas Wilson of Dundas, Ontario. A bottle-shaped burner, 28 feet in diameter, 125 feet high, and covered with a 15-foot-high wire hood, erected at a cost of $25,000, replaced the deep pit in which sawdust and other refuse from the mill had previously been burned.

The building was fitted up with the latest styles of saws and machinery, installed by Wm. Hamilton and Co., under the watchful eye of superintendent J.C. Else. There was one stock pony saw, a slabber, two gang saws, two twin cylinder saws, one large circular saw, and a lath mill. With this equipment, the mill produced an average of 150,000 board feet of lumber and 30,000 laths in an eleven-hour working day. On average this amount of lumber represented 1,200 logs. A machine shop, 60 feet by 80 feet, containing the most up-to-date equipment, ensured that the saws were kept in proper running order.

In 1883, as demand for lumber increased, the mill was enlarged, and a separate, smaller mill was erected in front of the main mill. Small logs were cut in this mill, reserving the main facility for large-diameter logs. Construction of the small mill and renovations in the main one increased cutting capacity to 300,000 board feet per day.

When market demand for lumber increased after 1880, the logging operations of Georgian Bay Lumber increased accordingly. Timber limits in Draper, Morrison, Muskoka, and Ryde townships in southern Muskoka which had remained virtually untouched during the depression were now harvested. From these far-away townships, logs had to be driven some twenty-five to thirty miles down the Severn River drainage to Georgian Bay and towed by tug to Waubaushene. Some logs, of course, were cut in the Port Severn mill. To facilitate passage of logs down streams and tributary rivers into the Severn, a network of dams, slides, and retaining booms had to be constructed.

While Georgian Bay Lumber was the largest operator in the Severn River watershed, it was not the only one, so a question arose as to who should build and maintain the dams and slides and what tolls, if any, users other than the builders of the dams and slides should pay. A similar situation existed on some 234 other rivers and streams in Ontario. The dilemma was resolved in March 1881, when the legislature passed the controversial Act for Protecting the Public

Mill engine room, Waubaushene c1890 – courtesy Robert Thiffault

Boilers, Waubaushene Mill – courtesy Robert Thiffault

Interest in Rivers, Streams and Creeks[14] and a companion piece of legislation, the Timber Slide Companies Act.[15]

The latter act permitted the formation of jointstock companies with authority to construct dams, slides, and other works to facilitate transmission of logs and timber down rivers and streams. The companies formed under the act could not prevent others from using their works, but they were empowered to charge tolls; these had to be approved by the lieutenant-governor-in-council.

Under authority of the act, William Earl established the Severn Driving and Boom Co., with share capital of $50,000, divided into 500 shares of $100 each. The charter allowed the company to build slides and dams along the Severn River at Big Chute, Six Mile Lake, Lost Channel, Ragged Rapids, and Macdonald's Rapids. More dams and slides were constructed on the tributary streams: on Morrison Creek into Morrison Lake in Wood township, on the Kahshe River into Kahshe Lake, and on Garter Snake and Riley creeks into the centre of Ryde township. The charter authorized construction of another series of dams and slides on the Little Black River (now Matchedash River), which drains Matchedash Lake in the north end of Orillia township through the centre of Matchedash township into MacLean Lake, above Port Severn.

Many of these dams and slides, particularly those on the main courses of the Severn and Matchedash rivers, had previously been built, but identifying them in the charter of incorporatiuon clarified their ownership and gave the company the indisputable right to charge tolls for their use. Tolls for passage of logs from Matchedash Lake to Port Severn were $1 per thousand board feet and from Garter Snake Creek to Port Severn $1.10 per thousand feet. Saw logs were estimated to contain 200 board feet. Each log had to be clearly stamped with the lumberman's mark, and the company had to be advised before 1 April of the number of logs in the spring drive and the destination. Sorting jacks were set up on Sparrow Lake and at Port Severn. The company's charter was to run for ten years, at the end of which time it was assumed that the estimated 230 million board feet of lumber in the Severn River watershed would be cleaned out. The Maganettewan Driving and Boom Co., with similar powers to improve the Still and Magnetawan rivers and tributary streams, was chartered in 1883. Creation of these two driving and boom companies was the last corporate act of Georgian Bay Lumber under William Earl's presidency. In fact, Maganettewan

Driving and Boom was chartered on 10 February, just six days before he died.

With the death of William Earl Dodge in 1883, the United States lost one of its shrewdest business entrepreneurs, and charities were deprived of one of their great benefactors. From the approximately $3.5 million his estate was estimated to be worth at the time of his death, he bequeathed $350,000 — a tithe — to various religious, educational, and cultural organizations. Canadians can be comforted with the knowledge that a portion of the profits generated through the slaughter of some of our magnificent pine forests went to support such worthy institutions as the American Bible Society ($10,000), the American Museum of Natural History ($5,000), the American Sunday School Union ($10,000), the American University of Beirut ($20,000), the Children's Aid Society ($5,000), Lincoln, Howard and Atlanta Universities ($10,000, $5,000 and $5,000, respectively), the Metropolitan Museum of Art ($5,000), the YMCA ($5,000) and several other similar organizations.

During the last ten years of his life, William Earl reorganized and strengthened three lumber companies on the shores of Georgian Bay that he had never intended to own. The work consumed much of his energy and a good deal of his capital, but he left his sons and heirs a well-managed, highly productive, and potentially profitable enterprise. What they would do with it would depend on how much of his tough-minded and hard-driving entrepreneurial spirit they inherited from him, along with the $250,000 he left to each of them.

Waubaushene c1890   – courtesy C.A. Stocking

Chapter Ten

The Sons Consolidate

The first act of William Earl's heirs was consolidation of the three lumber companies, with their five mills, into one — Georgian Bay Consolidated Lumber Co. Ltd. It did not make much sense to maintain the legal fiction of three independent companies, when in practice they were operated as one: each company had the same president, secretary treasurer, general manager, shareholders, and directors. Waubaushene was the head office of all three companies; the books for each were kept there.

The heirs wasted no time in initiating the process of consolidation; a petition to the lieutenant-governor for supplementary letters patent to increase the capital and change the name of the Georgian Bay Lumber Co. was dated 22 February 1883, only six days after William Earl's death. Supplementary letters patent, which increased the capital stock from $400,000 to $1 million and changed the name to Georgian Bay Consolidated Lumber Co. Ltd, were issued on 21 June. To effect the consolidation, the directors increased the share capital of the Collingwood and Maganettewan lumber companies from $150,000 and $300,000 to $200,000 and $400,000, respectively. Each shareholder received three shares in these companies for every two previously held.

Next, at a special directors' meeting, held in Waubaushene on 25 February 1885, the directors of each company (who were the same men) passed resolutions selling the mills and assets of the Collingwood and Maganettewan companies to Georgian Bay Consolidated Lumber for an amount equal to the revised capital value of each company. The directors then passed a pair of resolutions dissolving each of the companies and surrendering the charters.

Each shareholder then received one share of Georgian Bay Consolidated Lumber stock for each share held in the former companies. As a result of the complicated shuffle, the original capital value of the three companies had been increased by $150,000 — from $850,000 to $1 million — with proportionate increases going to each shareholder, but no new cash had been paid into the treasury of the consolidated company.

The provincial secretary found the method of liquidating the companies and transferring the assets highly irregular, because the provisions of the Ontario Winding Up Act had not been followed. However, because the shareholders in the companies were members of the same family, he did not pursue the legal question and allowed the reorganization to stand.

Arthur Murray Dodge was elected president, with an annual salary of $10,000. Cleveland H. Dodge was made general manager. Scott was made secretary treasurer, and Peckham was demoted to superintendent. Scott and Peckham were each given five shares but were not made directors. The directors were Arthur Murray, Cleveland, and William Earl Jr. Later Stuart Dodge, who acquired one thousand shares from his father's estate, became a director as well.

Arthur Murray's oldest brother, William Earl Jr — the "most conservative man in New York" — took five shares in the new company. There were two reasons for this. First, much of William Earl Sr's estate was tied up in the Canadian companies. When he died all the stock and outstanding debts he had in the three companies were taken over by Georgian Bay Consolidated Lumber. As principal executor of his father's estate, William decided to take a seat on the board of directors to ensure that the estate's assets were properly managed; but to be a director under Ontario law one had to be a fully paid-up shareholder.

Second, and related, Arthur Murray was only twenty-nine years of age, and William's son, Cleveland, was only twenty-three, when they took over control of this large investment; neither had much management experience. For the next several years, therefore, as a result of William's somewhat patronizing involvement, the directors managed Georgian Bay Consolidated Lumber with the interests of the estate in mind as much as the interests of Arthur Murray, for whom the company was the principal means of livelihood.

Annual board meetings were held in New York City in January or February, either in the Down Town Club or in Arthur Murray's house, at 72 East 34th Street. Two of the directors —

William and Stuart — were also executors of the estate. Cleveland acted as secretary. As president of the company, Arthur Murray chaired the meetings, which usually began with him presenting a balance sheet and annual statement for the preceding year. Plans for the coming year were outlined, questions answered, motions passed, and finally dividends declared. Although the board consisted of three brothers and a nephew, its meetings were always formal and conducted in a business-like manner, with discussion and major decisions properly recorded.

William generally asked the most questions. Although lacking first-hand knowledge of the lumber business, like Arthur and Cleveland, he certainly knew how to analyse a financial statement. Arthur faced such probing questions as the following:[1]

William: Why do you need so many managers?

Arthur: Since 1886, I have reduced management costs by some $6,200 by not replacing Peckham when he resigned and rearranging the duties of other managers.

William: Why do you need so many tugs, and why do they cost so much to maintain?

Arthur: The company has four tugs. Total operating and maintenance costs range from $10,000 to $12,500 a year. This seems excessive, but owning our own tugs, which are needed for towing booms of logs and bargesful of lumber, places the company in an independent position. Moreover, if one credits up the amount of work the tugs do at a fair value, they pay a good interest on investment.

William: Why do you have so much of the company's money at the credit of various employees?

Arthur: I intend to reduce this.

William: How are funds for the operation of the company obtained during the president's absence from Waubaushene for much of the year?

Arthur: There are two accounts in the Bank of Montreal — a general account and a special account. Cheques can be drawn on the general account only on Scott's signature, countersigned by me. The special account was created to meet emergencies and for the current business of the company. Scott, as treasurer, can draw on this account, but for all amounts drawn, a voucher has to be filed in the company's office in Waubaushene.

William asked searching questions and made frequent suggestions to ensure that costs were controlled and profits maximized, especially for the benefit of the estate. which held 4,965 shares in the company.

At the time of his death, in 1883, William Earl Sr's investment in the Canadian companies totalled $923,510.89 (see Appendix B), making up a large portion of his estate. The Canadian assets consisted of $482,400 in capital stock in the three companies, $374,444.23 in accounts due from advances that William Earl had made to the companies, and $66,666.66 outstanding on mortgages held with the Longford and Parry Sound lumber companies. Except for the latter amount, which would be paid in due course by the mortgagees, all the liabilities and the stock had been transferred to Georgian Bay Consolidated Lumber at the reorganization.

Because, by the terms of the will, the estate did not have to be settled for five years, the executors were willing to leave the estate's stock in that company temporarily, but in the mean time they needed considerable cash to pay some of the bequests. Mrs Dodge had been left the mansion on Madison Avenue, the country seat at Tarrytown, the horses, and all the personal property. These had to be maintained. In addition, the widow was to receive a lump sum of $800,000, on which she was to be paid an income of 6 per cent per year until the estate was finally settled. Each grandchild received $25,000 in cash, and there were many other smaller bequests, in addition to the large amount left to charities. One source of cash to pay these bequests was the outstanding debts owed by Georgian Bay Consolidated Lumber, but the company could not pay these without liquidating a good deal of its assets. An alternative to liquidation was sought.

At a special meeting of the board of directors held in New York in December 1885, a by-law was approved authorizing the directors to borrow $475,000 through the issuance of $100, 6 per cent bonds, against the credit of Georgian Bay Consolidated Lumber. Three hundred thousand dollars from the proceeds was to go to Mrs Dodge, and $175,000 was to be paid into the estate. When

James Scott learned about the by-law, he informed Arthur that it was illegal, because Canadian law prohibited the sale of bonds to pay existing debts. Arthur informed William, who turned Scott's letter over to the estate's lawyer, John E. Parsons of New York, for an opinion. Parsons agreed with Scott's interpretation and advised against proceeding with the bond sale. There was danger, Parsons advised, that if the company should become involved in debt to other parties before the bonds were paid, those other parties could challenge the validity of the bonds, and the security of the advances to Mrs Dodge and the estate would be in jeopardy. Canadian lawyer McCarthy was then asked for an opinion. He concluded that the transaction would be legal, because the charter of incorporation authorized the sale of bonds, debentures, and other securities, without limitation.

The matter of questionable legality notwithstanding, at a meeting in Toronto on 19 January 1886 the directors decided to proceed with the bond issue. Bonds payable in ten years, but redeemable at anytime with payment of three months' interest, were issued. To secure payment of the bonds and interest thereon, the directors issued to the executors of the estate a mortgage and trust deed on the mills and timber limits.

Later that year, the executors of the estate expressed the wish that all the Canadian interests be sold, if a suitable buyer could be found. Naturally, the directors of Georgian Bay Consolidated Lumber concurred. William was most anxious to sell, so the estate could be settled on schedule. He was probably not very happy about the bond sale either, because of its doubtful legality, and doubtlessly wanted the bonds redeemed as soon as possible. Moreover, he had never been very sympathetic to the Georgian Bay operations, because of the large amount of his father's capital that had been absorbed in bailing Anson out of his financial difficulties.

Cleveland Dodge, who held $125,000 worth of stock in the company, was anxious to redeem his holdings. Cleveland had originally intended to pursue a career in lumbering and, in fact, had served briefly as general manager of Georgian Bay Consolidated Lumber. But in September 1885, his older brother William Earl 111 died of diphtheria, and Cleveland gave up active participation in the lumber business to replace his brother as heir apparent in Phelps Dodge and Co. Having been married in 1883, he could now find other uses for his capital.

Arthur had the least motive for selling, because the company was his principal interest and primary means of income. However, since the estate, his brother Stuart, and his nephew Cleveland owned in total about two-thirds of the stock, he had no option but to agree to the sale. But one senses he did so reluctantly. With the redemption of his $327,000 worth of stock added to the $250,000 due him from his father's estate, he would be able to invest in something else.[2]

The task of finding a buyer was turned over to a real estate agent, who came forth with only one offer; it was made by the firm of N. Bradley and Sons, of Bay City, Michigan. Bradley made a conditional offer of $1.5 million for the mills, timber limits, and stocking (saw logs in inventory). He offered to pay $500,000 in cash and the balance with interest over five years. It was a good offer, but Arthur advised against accepting it. He had discovered that Bradley and Sons was worth only about half of what it was willing to pay for the company, and it was short of cash. Bradley Sr "appeared to be the backbone of Bradley and Sons, and as he was a man along in years, his lease on life could not be depended upon," and "his Sons & nephews did not make a favourable impression, as to their ability to manage so large a property."[3] Arthur expressed the further opinion that, since the balance statement of the company indicated a fair criterion of future prospects, the stockholders could probably receive much larger returns by working the property themselves than by selling it. The directors accepted his arguments and posponed the decision to sell.

Arthur was right. From a purely business point of view, it made no sense to sell the company. The five mills had a combined cutting capacity of 55 to 60 million board feet of lumber per year. Total production of sawn lumber in 1886 had been 52 million feet,[4] which produced a profit of over $2 per thousand board feet. The square-timber trade was also picking up. In 1886, 210,000 cubic feet had been sold, and there were orders for more. Because square-timber produced much higher profits than sawn lumber, averaging the equivalent of $13 per thousand board feet, the company decided to expand its square-timber production.

The company's books showed a profit of $147,169.22 for the year ending 1886, which, added to a carry-over of $42,064.52 for 1885, made a grand total of $189,233.52 in the profit and loss account. Consequently, a dividend of 18 per cent was declared as of 1 January 1887. The year 1887 was even more profitable, the year-end statement showing a profit of $209,172.52. With anticipated earnings for 1888, the board was able to declare a dividend of 50 per cent ($500,000) on the capital stock of the company. It would have been foolish to sell such a profitable enterprise.

With a decision not to sell having been made, Arthur then concentrated his energies on managing and expanding the company. In doing so he relied on the assistance and sometimes the wise counsel of his very competent Canadian managers.

It is not clear who suggested building a box factory, but the idea was a good one. Prior to erection of the box factory, edgings, slabs, and extremely coarse grades of lumber had been used for making lath, a large percentage of this being wasted. Also, short slabs and pieces of wood had gone into the burner, there being no way of using this material. The combined profits of the Waubaushene and Port Severn lath mills had been averaging between $8,000 and $9,000 per year, but there were indications that the price of lath was about to decline. Consequently, another and better way of using the waste material was sought; it was then that someone proposed using the material for making boxes for the American export market, principally for shipping coal oil.

In the spring of 1886, the box factory was built, for about $18,000, with Arthur Dodge and Delos Bliss, a partner in the Dodge and Bliss Co. (formerly Dodge and Meigs) of New Jersey, putting up the capital. (Later, construction costs were transferred to the books of Georgian Bay Consolidated Lumber.) The Dodge and Bliss Box Co. of Waubaushene, with share capital of $25,000, was chartered later in the year.

The box factory, which was located immediately behind the mill, and Georgian Bay Consolidated Lumber functioned in a symbiotic relationship. The mill sold to the factory scrap that had formerly been burned or converted into laths,[5] while the factory made substantial profits by manufacturing this waste into box shooks. In 1887, after an unprofitable first year, the factory netted close to $9,000 profit, while the mill made a clear profit of $3,000 from the sale of scrap. A similar but smaller box factory was built at Port Severn. The principal outlets for the shooks were Dodge and Bliss in New Jersey and A.M. Dodge and Co. of Tonawanda, New York. In time the boxes became very popular, and the demand was great. Installed by the Wm. Hamilton Co. of Peterborough, the machinery had capacity for producing 5,000 standard-sized boxes

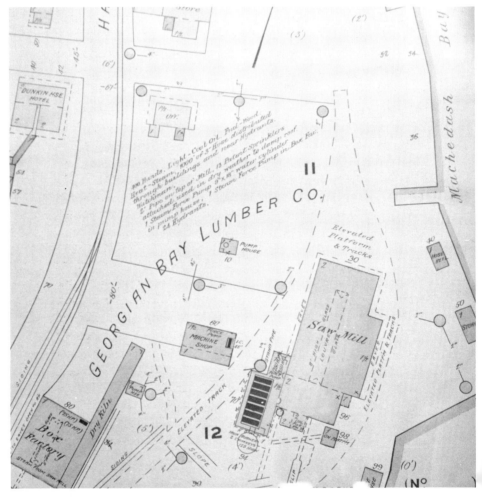

Plan of Waubaushene Mill 1890   – Simcoe County Archives

per day. The factory employed thirty-five hands, who turned out twenty-five to thirty different-sized types of boxes annually. By 1890, the factory was operating at full capacity to keep up with US demand for box shooks.

One of the unanticipated consequences of constructing the box factory was the fire hazard produced by the large number of sparks and burning embers spewed out of the burner from the light, dry refuse fed into it from the factory. Indeed, in May 1888, the roof of the office was ignited by sparks thrown from the burner; the fire was extinquished before any serious damage resulted. A new burner was required, and so one with a stack 136 feet high was installed in the spring of 1889, at great cost to Georgian Bay Consolidated Lumber. But despite the new burner, in October 1891 the Dodge and Bliss Box factory at Waubaushene met the fate of many similar, highly inflammable structures; at one o'clock in the morning of the twenty-ninth, the factory caught fire and was totally destroyed. The loss was $25,000. Insurance covered $15,000 of the loss and $5,000 worth of salvage was recovered, but the factory was never rebuilt.

Because of its remote location and great distance from world markets, the Georgian Bay Lumber Co. always had difficulty marketing its production of sawn lumber and square-timber. Principally, the lumber was sold to wholesalers in the United States and Canada, while square and waney timber was bought by wholesalers in Britain. Prime-quality lumber went to the United States; coarser grades were sold in the domestic market which, paid a much better price for it than the export trade. Wholesalers' agents from the three market areas visited Waubaushene every year and purchased quantities of lumber direct from the lumber yard. Because of the large quantity of lumber exported to the United States, the American government placed a consul in Waubaushene, who resided for several months of the year in the Dunkin House.

Transportation of lumber was always costly. As indicated earlier, before 1875, nearly all the lumber destined to US markets was shipped by barge or schoooner, while square-timber for Britain and lumber for the Toronto market were carried out on the Northern Railway from Collingwood. After 1875, much lumber was shipped on the Midland Railway which, in 1883, became part of the Grand Trunk system. In 1887, Georgian Bay Consolidated Lumber shipped some 27 million board feet of lumber by rail, and about 23 million feet by water.

Rail transportation was a constant source of frustration for the lumber company. It seems that there were never enough cars available to accommodate the tremendous production of lumber from the Georgian Bay mills,[6] making it difficult for the company to meet delivery commitments. Arthur explained the problem to the board of directors in his report of 1888: "One of the greatest difficulties we have had to contend with during the season has been that of the railroad, who have been unable to supply us with sufficient number of cars daily . . . We have been for days at a time with a full shipping crew, able to load 60 cars a day, when we could not get one-fifth of that number of cars; at the same time our expenses with this shipping crew had to go on."[7]

High freight rates were another bone of contention with the lumber company, especially after the Grand Trunk acquired ownership of many of the independent rail companies in Ontario, thereby gaining a virtual monopoly on rail transportation in the province. Arthur frequently tried to negotiate a reduction in freight rates with Joseph Hickson, general manager of the Grand Trunk, with lavish week-end parties in Arthur's luxury yacht *Skylark* providing a setting for the talks. But despite the convivial atmosphere, the talks were never successful. Water transport was cheaper than rail, but it was not always available either, because many of the shipping companies diverted schooners to carry wheat from burgeoning granaries in the west.

Although the company relied substantially on independent wholesalers and brokers for disposal of lumber, it also operated its own wholesale business. As discussed in chapter 3, Anson had formed partnerships with White and Co. of Albany and other wholesalers in Chicago and Cleveland to ensure distribution of his product. These outlets disappeared with the bankruptcies in 1873. William Earl sold some of his lumber through Dodge and Meigs, but he established outlets for Georgian Bay Lumber in Oswego and Syracuse that sold lumber to retailers in the northern states.

In 1883, the state of New York made the Erie Canal free from tolls; the Welland Canal continued to levy them. Consequently, Tonawanda and Buffalo, at the entrance to the Erie Canal, replaced Oswego as the receiving port for Ontario lumber. Arthur decided to set up a wholesale and retail lumber operation in Tonawanda. In 1886, with capital borrowed from Georgian Bay Consolidated Lumber, he bought several acres of land at Tonawanda on the Erie Canal, where he established a lumber yard, planing mill, and dry kiln. He then incorporated his own wholesale and retail business, A.M. Dodge and Co. The Oswego and Syracuse businesses were closed out. During the first year of operation of A.M. Dodge and Co.,12 million board feet of lumber — about one-quarter of the year's production of Georgian Bay Consolidated Lumber — was placed in the hands of A.M. Dodge and Co., which sold it on commission. Arthur raised the delicate question of an appropriate commission with the board of directors at the 1887 annual meeting. The board agree to pay A.M. Dodge and Co. 12.5 per cent, an amount slightly higher than the trade average, but in return Arthur agreed to handle several grades of lumber and to run the lumber through his planing mill, enabling Georgian Bay Consolidated Lumber to dispose of large quantities of stock at much higher prices than they could otherwise receive.

The Tonawanda agreement proved so successful that the directors decided to place as much of the production of the Waubaushene and Port Severn mills with A.M. Dodge and Co. as possible. Because of the increased expense incurred by that firm in handling a large volume of lumber and running it through the planing mill and kiln, Arthur asked for and received an increase in commission to 15 per cent. Although A.M. Dodge and Co. obviously acquired larger profits from this arrangement, Georgian Bay Consolidated Lumber benefited, too. At the end of the season, unsold lumber — sometimes very large quantities of it — was stored in the lumber yards at Port Severn and Waubaushene. Bank interest on capital tied up in these inventories cut into profits. By shipping this lumber to Tonawanda on consignment, where it was seasoned in the dry kiln, indirect storage costs were reduced.

And so at the end of the 1880s, Arthur Dodge, like many other large American lumber producers, acted in four capacities: through his timber licences, he was an owner of raw material; he was a manufacturer; and through A.M. Dodge and Co. he was both a wholesaler and a retailer of lumber.

The Georgian Bay Lumber Co. had, since the beginning of its existence, operated stores at each of its mills. The stores were essential for providing food, material, and equipment for the logging camps and river drives and for the inhabitants of the mill villages. In 1887, the company operated four stores: two at Waubaushene, one at Byng Inlet, and another at Port Severn. The stores were a highly profitable branch of the business: much of the money paid out to the workers returned to the company through its stores. The financial statement of 1887 showed a net profit in the stores' account of $8,885.94, on an investment of some $50,000.

But there were problems in the operation of the stores. Before 1886, each store operated more or less independently, with each storekeeper responsible for purchasing and delivering supplies to the associated logging camps. This arrangement increased costs and in some cases allowed petty thievery, or, as Arthur called it, "unnecessary leakage."[8] In 1886, management of the stores was concentrated in Waubaushene, with purchase and distribution emanating from that point. With the stores department under the control and inspection of head office, large savings resulted, but the extra bookkeeping placed a heavy burden on James Scott's clerical staff. When Scott decided to retire in 1887, the full burden and responsibility for the company's financial affairs fell on Arthur. With a view to eliminating some of the bookkeeping burden, Arthur recommended to the board of directors "the advisability of forming a stock company of [the] store business — a wheel within a wheel."[9]

It would be an independent company, but under Arthur's overall management. Georgian Bay Consolidated Lumber would control general policy, credits, amounts of purchases, prices and so on. The stores company would keep its own books, relieving the Waubaushene office of all accounts and all statements that did not directly belong in the lumbering operation. Arthur suggested W.H.F. Russell as an appropriate manager. Russell had been with Georgian Bay Lumber for about ten years, most of which time he had been in charge of the Waubaushene store. Arthur considered him quite capable of assuming responsibility for the new company. And if he were permitted to take up some of the stock, it "would not only give him special zest in his work but would tie him to us,"[10] Arthur advised the board.

The board accepted Arthur's suggestions, and in June 1888, letters patent were issued by the Dominion secretary of state, chartering Dominion Mercantile Co. Ltd., with share capital of $50,000 divided into 500 shares of $100 each. Half the stock was subscribed for upon incorporation. Arthur, with 173 shares, was the major shareholder. Russell and Joseph Hartman, keeper of the Port Severn store, each subscribed for five shares. Russell was made general manager, as planned.

It is curious that the new company was incorporated under Dominion, rather than provincial charter, the first of the Dodge

Village of Byng Inlet c1890 — courtesy Robert Thiffault

"The Store", Waubaushene c1890 — courtesy C.A. Stocking

companies to be so set up since Anson's lumber companies in 1871. Power granted to the company "to transact all business which is usually transacted by general merchants," and authority for "the operation of the said company . . . to be carried on in the Dominion of Canada,"[11] indicate the company's intention to broaden the range of goods carried in its stores, and perhaps Arthur's intention to enter general merchandising after the reserves of pine around Georgian Bay were exhausted in about twenty years. The company's charter clearly gave him authority to operate stores anywhere in Canada.

In June, a thirty-foot addition was made to the Waubaushene store to accommodate the full range of dry goods, clothing, shoes, hardware, glass, paint, oil, and food stuffs of all kinds. Large advertisements were placed weekly in all local newspapers. The outlets remained "company stores" of the classical mode, but, unlike some in one-industry villages, these did not gouge their customers. The Dodges, being Christian gentlemen, were too honourable to permit exploitation, and, being extremely munificent employers, they would hardly have overcharged their employees for the necessities of life. Moreover, the Waubaushene and Port Severn stores did not have true monopolies. Other stores were operated in Waubaushene by private individuals on private land, and competition from merchants in the nearby villages of Coldwater, Fesserton, Midland, and Victoria Harbour kept prices at Waubaushene and Port Severn in line.

However, one of the few mean actions ever taken by Georgian Bay Consolidated Lumber resulted from formation of the store company. For years, Georgian Bay Lumber had operated a pay office in Gravenhurst, where shanty men were paid when they emerged from the bush camps in the spring. With over one thousand men working in these camps, serviced from a depot in Gravenhurst, the amount paid out in wages each spring was great. In March 1888, for example, the Gravenhurst pay office distributed $10,000 in wages to jobbers and contractors alone. Naturally, merchants in Gravenhurst and Bracebridge benefited from this large injection of cash into the communities.

In 1889, the company decided not to open a pay office at Gravenhurst, forcing all the shanty men to travel to Waubaushene for their pay. The distance from the office where payments were made to the company store was only a few feet. There new clothing, tobacco, fruits, candy, and other luxuries, denied young men who had existed on salt pork and beans for six months, were laid out in tempting display. As might be expected, the men spent a good portion of their winter pay in Waubaushene, before departing for their homes in eastern Ontario or Quebec.

The merchants in Gravenhurst, who considered the logging industry in Muskoka their economic lifeblood, were incensed. From a business point of view, it made good sense for the company to pay the wood-cutters in Waubaushene, but it was a poor public relations venture. Russell, exercising "special zest in his work," may have been responsible for the decision to close the Gravenhurst pay office, but the Dodges would have approved it.

The Byng Inlet mills were a source of anxiety to the directors of the Georgian Bay Consolidated Lumber Co. Because of their out-of-the-way location, the mills' operating costs were higher than those of the other mills. In 1886, even though they had decided not to sell the whole company, the directors considered selling the Byng Inlet mills; consequently, for the next few years the mills were operated "on the principal [sic] of simply keeping the operation alive,"[12] and the mills cut only a quarter of the quantity of logs they were capable of cutting. A large numbers of sawlogs were sold to Michigan producers. The mill equipment had deteriorated badly, to the point where the Page mill had not even been operated for some time. By 1888, it would have been difficult to sell the mills, because of their run-down condition.

Still, there was some good-quality pine on the company's limits in the Wanapitei River area, of much better quality than the logs sawn in the Waubaushene and Port Severn mills. The Byng Inlet mills were closer to this pine than the others, and because the timber would stand a greater-than-normal manufacturing expense and still produce a profit, it was decided to upgrade the Byng Inlet mills. In January 1888, J.C. Else and boiler engineer D.B. Anderson, with one or two helpers, were sent up from Waubaushene through deep snow drifts to make the necessary repairs. Some $5,000 was spent increasing the engine power of the boilers and putting the saws in proper running order. In the spring, both mills were put into production, but the results "were so discouraging that the President felt constrained to ask the Directors to bear with this Department another year."[13] Then, if better results could not be obtained, Arthur advised that the mills be sold or closed down.

Results did improve. In the summer of 1889, the two mills produced 20 million board feet, and in the 1890 season the Dodge mills and a third mill, operated by the Burton Brothers of Barrie, cut an unprecedented 40 million board feet — the largest production of any mill town on the Georgian Bay.

Prospects for the Byng Inlet operation appeared good, but in August 1891 disaster struck: the Anson mill caught fire and was totally destroyed, along with a large quantity of lumber. Fortunately, the estimated loss of $85,000 was completely covered by insurance. The directors, wearing the hats of estate executors, then decided to sell the Page mill and all the timber limits, so that the bonds and stock held by William Earl's estate could be liquidated.

In November 1891, Arthur presented an offer to purchase from E.J. Lynn, a practical woodsman and lumberer from Detroit, who was backed by financier George H. Hammond, son of the former head of the Hammond meat-packing firm. Lynn and Hammond were prepared to pay as high as $600,000 — $200,000 in cash and the balance in three annual payments — if the property were found to be as valuable as represented. Because a business detective, hired by Arthur, reported the Hammonds to be worth about $20 million, and both Lynn and George Hammond to be trustworthy businessmen, Arthur was authorized to proceed with the sale. A minimum acceptance price of $500,000 was set by the board.

Arthur recommended that a timber berth adjacent to the Wanapitei River be sold as well. This berth, thirty-six square miles in area, bought by William Earl during the depression for only $8,000, contained about 125 million feet of excellent timber which, at current prices, was now worth at least $300,000. F.B. Chapin of Cartier, Quebec, an agent for the Michigan lumber firm Sibley and Barings, had requested a thirty-day option to buy the berth. Arthur recommended giving the option at a valuation of $300,000, but with a floor price of $200,000. The board accepted the recommendation.

Arthur hoped that the proceeds of the proposed sales, together with the insurance payments from the Anson mill and box factory fires, would enable him to pay into the company's treasury, in the next two or three years, an amount equal to the capital stock of the company. Also, in the course of general business, he expected to be able to pay in full the bonds and outstanding dividend notes, leaving the company in possession of the remaining three mills and timber limits, representing about two-thirds the value of the present holdings. This was encouraging news to William and the other executors who had been anxious to windup the estate.

Unfortunately, for reasons unknown, the Lynn/Hammond and the Sibley/Barings sales did not materialize. Another buyer was found in 1892, but by then the directors had decided to liquidate the whole of Georgian Bay Consolidated Lumber. That event will be discussed in chapter 12.

Arthur Murray Dodge 1852-1896

– courtesy Pauline Webel

Chapter Eleven

Arthur and Josephine

Arthur Murray Dodge was born in New York City on 29 October 1852, the youngest of the seven sons of William Earl and Melissa Dodge. One of three Dodge boys to go on to university from high school, Arthur attended Yale, graduating in 1874. After a customary interval of foreign travel, he entered the family lumber firm Dodge, Meigs and Co. of New Jersey, in March 1875. After serving an apprenticeship, like any ordinary clerk, during which he learned the various facets of the lumber trade, he was made a junior partner.

Also in 1875, Arthur married Josephine Marshall Jewell, second daughter of Hartford industrialist Marshal Jewell, whom Arthur had met in Europe. Josephine's father, a self-made millionaire like his own father, had been governor of Connecticut, and US minister to Russia and was, at the time of Josephine's wedding, postmaster general of the United States.

For the first few months of married life, Arthur and Josephine lived with Arthur's parents in the Madison Avenue mansion. Later, they moved into a rented mansion — 72 East 34th Street — in New York. They also maintained a summer home in nearby Riverside, from which Arthur commuted to work in New Jersey. The first three of their six sons were born in rapid succession: Marshall in 1876, Murray in 1878, and Douglas in 1879. After an interval of six years, the remaining three were born: Pliny in 1885, Geoffrey in 1886, and Percival in 1891. Pliny died in 1889 at the age of four.

Arthur and Josephine were well known in New York social circles. They entertained lavishly and often. They were also recognized for their untiring efforts in support of many charitable organizations, the two activities — charity and entertainment — often tied together. Their grand home on 34th Street was judged by one columnist to be "as well suited for a ball as any in New York."[1] The balls held there were large and, reflecting Arthur's and Josephine's personalities, always festive. The musicales which Josephine presented in the gold ballroom in support of charities were often attended by as many as 400 guests.

Arthur was a stout supporter of Josephine's many charity fund-raising activities. He also became a charter founder and first treasurer of the Charity Organization Society of New York, a forerunner of Community Chest. But Arthur's principal interests outside business were sports and outdoor recreational activities. He was a shareholder in the North Carolina Hunting Club, held memberships in the Rockaway Steeple Chase Association and the New York Riding Club, and was a member and strong supporter of the Yale University Club. An avid yachtsman, he was "honorary skipper" of his own steamer yacht, *Skylark*. He also owned two small sailboats, *Ida* and *Jewell*, in which he enjoyed sailing on Georgian Bay with his three oldest sons.

Arthur was described by a contemporary as one with a "bright and active temperament . . . sincere and kindly nature and a frank and genial manner."[2] The New York *Tribune* called him "generous, joyous, hospitable, affectionate,"[3] while the Yale *Obituary Record* described him as "a man of warm sympathies and upright life, a generous friend of the University."[4] The Reverend Mr W.A. Wylie, Presbyterian minister at Waubaushene, expressed the feelings of the entire village when he paid the following tribute to Arthur: "A.M. Dodge [will] ever be held in loving memory by all who [have] ever felt the magnetism of his genial presence."[5] Wylie praised Arthur's "liberality and generosity" and his "countless efforts to promote the interests of the community."[6] Percy Dodge, who was only five when Arthur died, remembered that his father "was loved for his sense of humor."[7]

Although practising Christians, Arthur and Josephine did not live puritanical lives like the older Dodges. Arthur liked a pipe, and Josephine smoked cigarettes. According to her son Percy, Josephine learned to smoke and drink vodka in the court of the tsar during the three years she lived there, when her father was minister to Russia. "She was one of the first women in New York to smoke and the last one to do so in public,"[8] Percy observed. Percy Dodge has also given us a clear insight into his father's religious observance." He took the extreme nature of Wm.E.'s religion seriously, but never allowed it to overpower him,"[9] Percy wrote to Eugenia Price, after reading her portrayal of Arthur in *The Beloved Invader.*

Arthur kept a fine wine-cellar in Waubaushene, even when temperance was both a company rule and a way of life. On at least one occasion, violation of the company's temperance rule by the president had a humorous outcome. Arthur and Josephine frequently entertained senior officers in their home in Waubaushene or on *Skylark.* At one such dinner, the cook served brandied cherries for dessert. Mill superintendent J.C. Else, a staunch teetotaller, praised the dessert, which he had consumed with relish, and when asked if he cared for more replied, "No thank you. But I would like a little more of the juice."[10]

Arthur became the third president of the Georgian Bay Lumber Co. after his father died in 1883. During the several summers they spent at Waubaushene, Arthur and Josephine tried to breath a spirit of gaiety into the somewhat stuffy social atmosphere that existed under the sober-minded managers and superintendents appointed by William Earl. They were not completely successful. During Arthur's presidency, Waubaushene, and to a lesser extent Port Severn and Byng Inlet, evolved from primary-resource, bunkhouse communities into normal societies, with the same institutional structures found in any other village of similar size: family housing, schools, stores, recreational facilities, and community, church, and fraternal organizations.

One of the first management decisions made by Arthur, after taking over the presidency, was to employ married men in the mills, because married men had proved more satisfactory workers than bachelors. This change in employment policy profoundly affected the sex distribution and social structures of Port Severn and Waubaushene. With a large number of families taking up permanent residence, both villages grew rapidly and the populations took on greater stability. As a result of the enlargement of the Waubaushene mill in the winter of 1882-83, requiring an increase in the work-force, the population of Waubaushene grew even faster than Port Severn's, where the size of the work-force remained fairly constant.

To accommodate the growth, a large number of new houses had to be constructed; in the spring of 1883, fifteen were erected in Waubaushene. These cosy frame-houses — provided free of charge to married workers — were erected along well-laid out streets, beautified with small willow trees. Each house had a yard large enough for a garden and, following the practice initiated by Theodore Buck, was encircled with a whitewashed picket fence. One and a half miles of wooden sidewalks were laid that spring — some new, some replacing existing walks — providing greater accommodation and adding to the neat appearance of the village.

Waubaushene was a pleasant place. The summer scene, from sunrise, when the mill began operating, until sunset, was bright and animated. With its attractive homes, community hall, school, commodious churches with eloquent preachers, railway cars constantly entering and leaving the lumber yard, and numerous vessels loading and unloading at the wharf, the village exuded an aura of enterprise and progress. In May 1884, a reporter for the Orillia *Times* was inspired to write: "The beauty of the scenery, the advantages of bathing, the pure bracing air from the bay unite in conducing to the health and happines of all those whose privilege it is to have a home or spend a summer in the charming village of Waubaushene."[11]

Whether inspired by the charm of Waubaushene or simply conforming to the new social order, general manager Jesse Peckham, although a forty-two-year old bachelor, deemed it prudent to acquire a wife. Arthur and Josephine presented the couple with a fine twelve-place dinner set[12] as a wedding present, and Arthur authorized construction of a new home for Peckham, who, since taking over the general managership, had lived in the old house built by Theodore Buck.

Because it was a one-industry, "company town," no formal municipal organization was allowed to develop in Waubaushene. With the exception of the small lot deeded to the Roman Catholic diocese and the school property sold to the school board in 1882, the company owned all the public buildings or the land on which they stood and nearly all the houses; therefore, all major decisions regarding village planning and improvements were made by company officers under Arthur's direction. Even civil justice was dispensed, when necessary, by the company, Secretary Treasurer James Scott having been appointed justice of the peace by the Ontario government. But if the company made all the major public improvement decisions, social and recreational activities were organized by the residents themselves, usually under the leadership of elected committees.

Each summer, the churches organized family picnics. On Dominion Day (1 July), the mills at Waubaushene and Port Severn were closed down for the day and regattas and public picnics were held. At Waubaushene, races of all kinds were organized, the double scull competition being one of the more popular. These regattas were not limited to Waubaushene residents; contestants came from other communities. Prize money was subscribed by the leading citizens and merchants, the company putting up a large portion. In July 1884, the first year the Dominion Day festivities were organized, the large sum of $225 was collected for prizes at Waubaushene. Port Severn's celebration that year consisted of a mammoth public picnic on one of the islands in Georgian Bay; nearly the whole village was transported there on the steam tug *H.L. Lovering*. A large dance floor, erected on the island, allowed the adults to end the pleasant day with dancing under the stars.

Regular social, cultural, and recreational activities took place during the winter months. The company donated a piece of land on which a hockey and curling rink was built by private subscription. Those teenagers and young men who did not go off to work in the logging camps organized a hockey team which competed with teams in neighbouring communities. Through the years, Waubaushene produced many fine hockey players, some of whom acquired prominence; one or two played in the National Hockey Leauge.

In the mid-1880s, a curling club was organized, with Arthur and Josephine allowing their names to stand as patrons. J.C. Else was president; the management committee consisted of J.C. Arnold, G.F. Hall, and C.P. Stocking. In 1890, the club advanced to the Ontario Tankard bonspiel in Barrie but was not successful.

Each winter a carnival was held, usually in late February. These featured a brass band from Coldwater, a costume parade, skating races, dog-team races, and ice-boat competitions, extremely popular in the 1890s and early 1900s. (C.P. Stocking and D.B. Anderson built what was probably the first ice-boat in Waubaushene, in the winter of 1890.)

Waubaushene had a fine orchestra which held regular concerts in the community hall and played for dances in the rink. The churches also held concerts in the winter, and travelling theatre groups occasionally performed in Waubaushene. As her first contribution to community life, Josephine Dodge presented to the village an attractive library reading room, with many of the books provided from her own home in New York and the collections of her friends. The library, like the Union Memorial Church donated by William Earl and Arthur — both of which are still standing — was lined throughout with local ash.

A branch of the Independent Order of Odd Fellows (IOOF), still active in Waubaushene, was organized in 1879. Most of the senior officers and captains of the company and some independent businessmen were members. With volunteer labour and materials donated by the company, the members built a fine hall on land also donated by the company. The Reverend Mr R.J.M. Glassford, minister of the Presbyterian church, was chaplain for many years. Annual balls put on by the Rebekahs — the female arm of the IOOF — were social highlights in the village. In later years a chapter of the Imperial Order Daughters of the Empire (IODE) was formed, as well as a branch of the Women's Institute.

Outside the churches, which almost everyone attended, the most common social gatherings were house parties. Activities organized by the local elite were frequently reported in the social columns of the *Coldwater Tribune and Waubaushene Investigator* and the Orillia *Times and Expositor*. Parties usually featured live music and "innocent amusements,"[13] including the popular parlour game Proverbs, a contest in which the participants were required to identify proverbs from clues provided.

When Arthur Dodge died in 1896, the company took over maintenance (but not ownership) of his mansion. A social club, with headquarters there, was formed, to organize a weekly dance. The recently appointed general manager, W.J. Sheppard, an infrequent attender, was named honorary president. The active officers were J.C. Else, president; Charles P. Stocking, vice-president; Charlie Sheppard, secretary treasurer; and C. P. Wallace, master of ceremonies. The house was also used for private parties and weddings.

Formation of Georgian Bay Consolidated Lumber required some administrative changes. As indicated, Arthur was elected president, and his nephew, Cleveland H. Dodge, who had purchased 1,250 shares from his grandfather's estate, was appointed general manager. Jesse Peckham was no longer needed. He was demoted to superintendent, a position with little authority because the mill managers reported directly to Arthur and Cleveland.

Nepotism, long a hallmark of Georgian Bay Lumber, was continued by Jesse Peckham. He brought three of his relatives to Waubaushene — his brother-in-law and former partner Isaac Hoag, Hoag's son Fred, and Peckham's twenty-year-old nephew Charles P. Stocking, oldest son of his widowed sister, Caroline. Charlie Stocking had been serving as a clerk in the Newmarket Tannery. In

August 1886, as part of a cost-reducing measure encouraged by William Earl Jr, Peckham and his relatives, except young Stocking, were let go.

It is not clear whether Arthur requested Peckham's resignation or whether it was tendered voluntarily, but, in any event, Peckham resigned suddenly and left for Newmarket in the second week of August. The Orillia *Times and Expositor* reported "unsatisfactory health" as "the principal cause of his resignation,"[14] but the Newmarket *Era* the next day observed only that "Mr Jesse Peckham has retired from active connection with the lumber company and taken up residence in Newmarket."[15] There was no mention of a health problem. Immediately after returning to Newmarket Peckham rejoined the Newmarket Tannery Co.

Arthur informed the board of directors of Georgian Bay Consolidated Lumber that "a saving of $6200 as against the salaried officers beginning in the year 1886,"[16] had resulted from the resignation of Peckham and others. Treasurer James Scott and experienced lumbermen William J. Sheppard, Henry L. Lovering, and Joseph Hartman were kept on, at combined salaries of $8,500.

Arthur had been preparing for the change in management for some time. In 1882 a second mansion was built just across the road from the former Buck House, then occupied by Jesse Peckham. That summer Arthur transferred into his own name 28¼ acres of company property on the hill containing the two mansions and a stable. The following year, when it was decided that Cleveland was to become general manager, Arthur sold him an undivided half of the Dodge compound for $5,926. Cleveland and his bride were to occupy the Buck House during summer working vacations, and Arthur and his family took over the new mansion. Peckham, having been demoted to superintendent, was not entitled to one of the mansions and in July 1884 moved into his new, smaller house built at the top of Willow Street.

Cleveland Dodge actually spent two summers at Waubaushene — 1883 and 1885. He did not visit Wabaushene in the summer of 1884 because his wife was expecting their first child. Then, as stated in the last chapter, he gave up his position in Georgian Bay Consolidated Lumber to replace his deceased brother in Phelps Dodge and Co., and Arthur took over sole management. After Peckham's resignation, Arthur informed the people of Waubaushene that he "intended to be very regular in his attendance [at Waubaushene] and personally superintend the business."[17]

True to his word, Arthur and his family spent the next ten summers in Waubaushene, and for two or three months of the year, when the Dodges were in residence, the village took on the attributes of a fief. To live there in the 1880s and 1890s was like living on a royal estate in Britain or perhaps in the tsar's summer palace complex at Peterhof. Like a royal family retreating to summer residence in the Highlands, the Dodge family arrived at Waubaushene in June on a special train from New York. The day of arrival had been announced well in advance so that villagers could turn out en masse to greet them.

Arthur and Josephine stepped down from the train first, followed by the excited young "princes," Marshall, Murray, Douglas, the toddler Pliny (until his death), and baby Geoffrey (succeeded in 1891 by Percival). With the boys was their private tutor, Mr Holister. Next came the dowager, Melissa, who, as principal beneficiary and chief executor of William Earl's estate, controlled the capital on which Arthur's lumber kingdom rested. Melissa was accompanied by her personal pastor, Reverend Dr McIlvanie, who conducted daily family prayers and served as guest preacher to the Presbyterian congregation. Following Melissa off the train came the courtiers — the many friends and relatives who would provide the social succour without which life in the summer palace for the party-loving Josephine would have been unbearable. Neither the Dodges nor their children mixed socially with the simple village folk. Finally came the host of house servants — cook, waiters, nurses, nannies, and Josephine's personal maid. Locals were hired as servants, too. Young Lizzie Riddle, daughter of the estate's caretaker, Thomas Riddle, was employed as a lady's maid in the summer of 1889 and had the added good fortune of being taken to New York on a permanent basis.

Arthur's summer residence at Waubaushene had neither the elegance nor the durability of Anson's Beechcroft, but it was a modern and functional structure. Had it survived, Waubaushene could now boast an architectural showpiece: one of Canada's few pure examples of Second Empire architecture with its distinguishing Mansard roof.

The practical-minded Dodges found the dual-pitched, hip roof style useful, because it provided additional space for a full third storey. Whoever designed Dodge House knew the style well and faithfully incorporated all its elements into the building: projecting dormer windows on the steep lower slope; roof facades decorated

The Dodge House, Waubaushene c1888   – courtesy C.A. Stocking

The parlour in the Dodge House, Waubaushene c1890
– courtesy Robert Thiffault

Drawing room, Dodge House, Waubaushene c1890
– courtesy Robert Thiffault

with elaborate cornices supported by decorative brackets. Typically, the building was symmetrical, but slightly longer than wide. To relieve the monotony of the solid but stately structure, a projecting central bay, topped with a low, flat tower, was built into the front of the house; second-floor windows were shuttered. A projecting porch surrounded the house on three sides.

Not surprising, the house was constructed entirely of wood. Cedar shingles covered the roof; horizontal cladding, characteristic of Second Empire architecture, consisted of narrow clapboard. The kitchen and utility rooms were attached to the back. In 1889, after Pliny and Geoffrey were born, an unmatching, two storey, gable roofed addition with vertical siding was tastelessly added. Many of the guests and undoubtedly some of the servants slept in the original Buck/Peckham house, located a few yards away.

Second-and third-floor rooms were grouped around central halls. The second floor was reached by a staircase leading from the drawing-room; twin stairways led to the third floor. A drawing-room, a dining-room, a parlour, and probably a library were on the main floor. The parlour was a showcase of classical elegance and cheerful modernity, reflecting the exquisite tastes for which Josephine was well known and revealing also her awareness of the late Victorian trend toward a less garish decor.

The room was a riot of gaily coloured flowers and leaves superimposed on pastel backgrounds: walls were covered with English floral wallpaper; matching damask drapes, suspended from heavy rods by solid brass rings, covered the windows; lamp shades were of flowered silk, and a straight-back lady's chair and matching footstools were upholstered in light fabrics, with floral patterns. The floor was covered with a large, magnificent, light-toned Persian carpet. The latest, white-painted Jacques and Hayes parlour furniture, manufactured in Toronto, predominated — settees, fern and tea tables, and matching footstools. The light furniture was offset with contrasting dark, antique Federal-style chairs scattered around the room. Exquisite silk-shaded Bradley and Hubbard-type table lamps and a brass filigree chandelier provided ample lighting. The room contained little of the bric-a-brac that typically cluttered Victorian parlours. Only a Chinese vase or two, family portraits, a music box, and a china jardiniere, with the ubiquitous fern plant, appear in a photograph taken around 1890. Portraits of relatives and landscapes, most in classical frames but a couple in rococo Victorian, were tastefully arranged on the walls.

The drawing-room was also amazingly uncluttered and showed evidence of Josephine's handiwork. The walls were panelled in vertical pine, with overlay moulding on the joints. The floor consisted of alternating planks of dark walnut and white ash, creating a laminated effect that stood in sharp contrast to the intricate designs of the exquisite Persian scatter carpets. The fireplace had an Adams wooden mantel with a characteristic band of hand-carved dentil work; a metal hood to catch sparks hung over a low firescreen. A wide walnut staircase led from the drawing-room to the second floor. The lower newel post was crowned with a unique antique gas night-light, later converted to electric.

The house was originally lit with acetylene gas and oil lamps, but electric fixtures were installed around 1888, when a steam generator in the mill's power-house began producing direct current to provide incadescent lighting in the mill, the office, and the Dodge houses. One or two street lamps were placed at strategic locations in the village at this time as well. In 1889, a water system was introduced to the village. A pump house erected behind the office fed water to twenty-four hydrants scattered throughout the mill, the lumber yard, and various other company buildings. Water was also piped into some of the principal houses, including, naturally, Dodge House; subsequently, indoor plumbing was installed, rendering obsolete the myriad chamber pots that had adorned the bedrooms.

Each spring the house and grounds were painstakingly prepared for the arrival of the summer occupants: the rooms were aired, windows washed, wallpapering and painting done inside when necessary; the outside was painted every two years, whether it needed doing or not. Caretaker Thomas Riddle manicured the spacious grounds; shrubs were trimmed or replaced; flowers were planted and sidewalks were repaired. The miles of picket fences that enclosed the field at the back of the estate and surrounded the main grounds, isolating them from the closely quartered houses of the workers and managers at the foot of the hill, were freshly whitewashed every spring.

Arthur engaged in the kinds of hobbies that the landed gentry normally pursued in the nineteenth century. He raised purebred Jersey cattle at Waubaushene. The elite herd was serviced by a purebred bull with the regal name of Albert George, a noble beast that tipped the scales at 1,865 pounds. In September 1889, Thomas Riddle entered the bull in the cattle show at the Canadian National Exhibition in Toronto and won third prize.

Albert George c1890   – courtesy Robert Thiffault

One of the Dodge boys with the family's pet goat c1888
– courtesy Robert Thiffault

Arthur also experimented with raising Dorset sheep. In 1888, he brought a small flock of purebreds to Waubaushene. The following year, he purchased a prize ram at Port Credit, Ontario. Despite the fact that Dorsets were reputed to have the unique capacity for breeding twice a year, sheep raising was not successful at Waubaushene. But, presumably, the ram thought the attempt worthwhile.

For the amusement of the children, there was a small game park on the estate. There were ponies in the stable for the boys and a carriage horse for Josephine to explore the country roads. The boys had a pet goat that pulled them around the village in a small wagon. They also had a two-masted sailboat — *Jewel* — in which, when they were small, they sailed around the protected waters of Waubaushene Bay; later they acquired a much larger boat, *Ida*, that took them into the open water of Georgian Bay proper. Favourite adventures for the boys were fishing-camping expeditions to Big Chute on the Severn River, one of the most picturesque spots in the district. Pickerel and bass abounded at Big Chute; so did mosquitoes which, on at least one occasion, drove the campers back to Waubaushene.

Arthur's most expensive toy was *Skylark*, his father's steamer yacht, which Arthur brought up from New York. Prior to acquiring *Skylark*, he fitted up the steam tug *Marshall and Murray* (named after his two oldest sons), in which he and his guests enjoyed sightseeing excursions on Georgian Bay. But *Marshall and Murray* was not suitable for this purpose. It was an inelegant workboat, needed almost daily for company business, and had overnight accommodation fit only for the crew.

Skylark was already old and in poor condition when Arthur brought her to Canada in the summer of 1887. Her exact age then is not known, but she was in his father's possession in 1878 when she was reported to have been lying at anchor in the Hudson River in front of the Dodge mansion at Tarrytown, New York, when William Earl and Melissa celebrated their fiftieth wedding anniversary. Soon after being brought to Waubaushene, the yacht was refitted, redecorated, and upholstered "with a most lavish expenditure."[18] Josephine was in charge of the renovations. When they were completed, *Skylark* was alleged to be "without doubt the most elegant yacht afloat."[19] It was said that "the decorative work surpass[ed] one's imagination of beauty and reflect[ed] great credit of the refined tastes of Mrs A.M. Dodge."[20] Another reporter called *Skylark* "as perfect and as handsome a craft as sails anywhere on

Arthur and Josephine Dodge and guests on the tug Marshall and Murray *c1885*
– Huronia Museum

The Skylark *c1890*
– courtesy J.C. Miller

Arthur Murray Dodge on the Skylark *c1890*

– courtesy Robert Thiffault

the Canadian waters."[21] When improvements were made to her engine, she was able to slide smoothly and quietly through the water at eighteen miles per hour, with what was described as "a peculiar fairy-like appearance."[22]

But, despite her fairy-like qualities, *Skylark* was not very seaworthy. Designed for the sheltered water of the Hudson River, she was ill-suited for the open stretches of Georgian Bay and more than once was forced back to Waubaushene when she encountered rough water. She was a sleek ninety-footer with a narrow beam; she rode low in the water. Her boiler and engine were located midship, providing good balance, but, with her two masts, she looked more like a small ketch than a steamboat; indeed, she may have been a converted sailing vessel. In the fall of 1889, after the family had returned to New York, the yacht was put in drydock in Owen Sound, where she was given a new stern and her spine was stiffened.

Although Arthur was nominally in command of *Skylark*, he had the good sense to engage the experienced Capt. Jack O'Donnell as sailing master. George Wembridge, manager of the Dunkin House, was employed as steward. With her fresh paint, uniformed captain and crew, banners fluttering in the breeze, bells ringing, and whistle sounding, she was truly a magnificent spectacle. When the weather was fair, Arthur and Josephine took parties of their guests on sightseeing trips on Georgian Bay, sometimes attending the horse races at Collingwood, sometimes enjoying fishing or merely visiting picturesque and "romantic spots"[23] on the Bay; the excursions were always arranged with the particular tastes of the guests in mind. Arthur usually returned to Waubaushene in November with a party of male companions to take on hunting trips in *Skylark*. Afterwards, she would be tied up for the winter in the current below the dam at Port Severn, where the water generally stayed open all year round.

The residents of Waubaushene, as would have been most nineteenth-century Canadians, were awed by the show of wealth and the symbols of power displayed by the Dodges. Consequently, Arthur exercised charismatic authority over the workers and managers that drew from them the type of allegiance that is normally bestowed on a beloved monarch. How well the villagers loved their "king"! He, his family, and his friends worshipped with them on those Sundays when they were not cruising on Georgian Bay. Although the family was Presbyterian, Arthur contributed generously to the stipends paid to the pastors of the Methodist and Anglican congregations as well as to the salary of his own Presbyterian minister.

Children loved him, because he donated ice-cream and cases of lemons — rare treats in the 1880s — for the Dominion Day celebrations. When the ice rink burned down in 1890, he donated a better site for a new rink, opposite the IOOF Hall and facing the Coldwater Road. He was generous with both praise and bonuses to the managers and frequently entertained them at dinner. He regarded James Scott almost as a father, promoted him to vice-president, and frequently received him as a guest in his home in New York.

Local reporters for the district newspapers always wrote glowingly of Arthur Dodge. The "liberal minded son"[24] of the late William Earl Dodge, a reporter for the *Coldwater Tribune* called him in 1887. "Everyone will be pleased when the worthy President of the Georgian Bay Lumber Company returns,"[25] the paper predicted in June 1889. His popularity seems to have remained undiminished through the years: "Mr. A.M. Dodge Esq. is daily expected from New York when our residents hope he will, as usual, brighten things up about us,"[26] the local paper announced in April 1890.

Construction of the box factory in 1887, creation of a dry kiln, and further improvements to the mill produced an increase in the workforce from 150 to 200 hands. This stimulated another building boom in Waubaushene. Eighteen more houses were built in the spring of 1889, an addition was put on the Dunkin House, the Dominion Mercantile Co. added a second store, and a new, larger school-house, with a room for senior students, was erected. Also that spring a new, higher burner was built.

The intense construction activity earned Waubaushene a reputation as the "Chicago of the North," but unlike the rapidly expanding Chicago to the south, the namesake's social structure was not weakened by the dramatic changes in its physical proportions. The puritanical cement poured into the communal framework by the founders was too strong to allow social decay. Temperance was still the order of the day, maintenance and beautification of houses remained a condition of occupancy, and the Protestant work ethic and the stern uncompromising creed of the dominant Presbyterian congregation continued to determine the mores and create the social sanctions that controlled the behaviour of the villagers.

A year after the building boom was completed, when the population had grown to over eighteen hundred souls, a roving reporter for the Orillia *Packet* was able to describe the village as follows: "While Waubaushene is a small town, it is one of the best in Ontario. The lawns and yards are pictures of neatness, the fences in their new coats of whitewash present a healthy appearance; the people are genial, industrious, orderly and quiet. Vendors of strong drink — like the Irishman's potatoes — are few and far between."[27]

Everyone in Waubaushene was not always "orderly and quiet." One afternoon in June 1888, a villager distinguished himself by maltreating his spouse on a public street. In the morning he had ordered his wife to prepare him a warm bath, and when she did not comply with the request he proceeded to thrash her. Because "the same program was gone through in the evening,"[28] the culprit was paraded before Justice of the Peace James Scott. Scott reprimanded the wifebeater and threatened him with a dressing of tar and feathers if he did it again. Josephine provided a more practical solution to the bath problem. With "accustomed generosity"[29] she donated funds for the erection of a public bath-house with eight compartments.

Waubaushene's society was, if anything, too rigid. There was not a great deal for young single men, who still comprised a good proportion of the population, to do. Young women were closely guarded and strictly chaperoned. Mrs W.J. Sheppard, who had a family of five young boys and one daughter, refused to live in Waubaushene or any of the other mill villages. Sheppard established a home and kept his family in the quiet farming village of Coldwater. When a young man in Waubaushene craved the favour of the opposite sex, he stole away to the pleasure houses in nearby Victoria Harbour, where such institutions were tolerated. But he always announced that he was "going to Coldwater."

Arthur and Josephine always tried to provide leadership that was compassionate and just. Arthur was certainly an enlightened manager. In June 1887, without any representation from the men, he decided to reduce the working day from 11 to 10½ hours and got into difficulty with the Georgian Bay Lumbermen's Association for doing so. His original intention was to shut down the mill at 5:30 p.m. every day, instead of at six, but because of the Association's objection he kept the mill running until six and gave the men Saturday afternoons off instead. The men liked that arrangement better, because it provided an opportunity to go fishing or hunting or to travel by train to Orillia or Midland (or Victoria Harbour) to fraternize with friends. Also, in the winter of 1889, when there was a slump in the lumber market requiring reduction in the production of sawlogs, Arthur ordered the company to find employment in the village for all the men who would otherwise have gone off to work in the bush camps. His generosity strengthened the workers' loyalty to their employer.

A public furore was created in the summer of 1889, when Arthur permitted the public reading room to be opened on Sunday afternoons, although no books were to be taken out on that day. Arthur hoped that opening the reading room would keep young men "who resort thither . . . from pursuing more questionable amusements on that day."[30] The "unco guid" in Waubaushene objected to the sacrilege. They believed that "the sanctity of the Sabbath should be kept inviolate," that opening the reading room was the "thin edge of the wedge which would in time open up a way for introducing other modes of recreation harmful in their character."[31]

Perhaps because Arthur's own modes of recreation were considered by some to be ostentatious if not "harmful in their character," he decided to move his summer residence away from Waubaushene to a place where the family could have more privacy. In 1892, he purchased Present Island from New York sports editor James Watson. This thirty-five-acre island, located one mile southeast of Beausoleil Island and ten miles from Waubaushene had been acquired by Watson from the crown in 1890. Known to the Ojibwa as Wan-a-kiwyn, meaning "present" — a name originating with the practice of Indians receiving gifts from fur traders before serious trading commenced on the island. The island had been part of the land granted to the Coldwater Indians when they sold their reserve in 1836. They had moved to Beausoleil Island in 1842 and lived there until 1856, when they surrendered Beausoleil and several other small islands, including Wan-a-kiwyn, and moved en masse to Christian Island, where their descendants live today. Arthur built a large cottage and a dock on Present Island, complete with tennis courts, and even a liquor still. Henceforth, when the family or guests arrived at Waubaushene or Penetang, they boarded *Skylark* and were taken directly to the island. Arthur's connection with Waubaushene, thereafter, was limited to the brief periods each week he worked in the office.

The Dodges and guests at Waubaushene c1890, Josephine third left, Arthur on the right — courtesy Pauline Webel

Dodge Cottage on Present Island c1900 — Huronia Museum

In February 1892, the Dunkin House burned to the ground. A replacement hotel would later be built, but because, in the mean time, single female employees such as school teachers and secretaries had no suitable place to live, the Buck/Peckham House, which was then not needed for housing family guests, was rented to the company and converted into a boarding-house. When Edward V11 ascended the throne in 1901, the boarding-house was named Alexandra House, in honour of the new king's consort.

When theGeorgian Bay Consolidated Lumber Co. was dissolved in 1893 and the possibility that Arthur would sell his interests in Canada existed, he had the village of Waubaushene surveyed into lots and a town plan drawn up to facilitate disposal of the property. While most of the village residential area was divided into small lots, his own personal estate was divided into three large blocks: block 1, at the west end, consisting of about two acres, was divided into 9 lots; block 2, in the centre, comprising 9.68 acres, contained his summer house; block 3, with 16.73 acres, contained the Buck/Peckham boarding-house.

As the next chapter will reveal, Arthur did not sell his lumber interest, but in 1895 he did sell block 3, with the boarding-house to the new Georgian Bay Lumber Co. for $3,500. He kept block 2, with his summer house, although he used it only rarely after the Present Island cottage was built, primarily when he visited Waubaushene in the winter and in the early spring and late fall, before and after his family joined him for summer vacation.

The village of Port Severn had been a matter of concern to Arthur and his predecessors for some time. Although the company owned the mill, the service buildings, and the houses, it held the land on which they stood only under lease from the crown. The lease, which covered about 109 acres, consisting of broken lots 19 in the twelfth and thirteenth concessions of Tay township and the main island of Port Severn, dated from 1830, when the crown set the property aside as a reserve for the Indian saw mill. When William B. Robinson bought the mill from the Indians, he was given a crown lease to the property. Through the years, the lease had been reassigned to successive purchasers of the mill. ending up finally in possession of Georgian Bay Consolidated Lumber. Unhappy with this arrangement, the company applied for and was eventually given a patent to the Port Severn mill property, in November 1881.

The security of the French-Canadian settlers at Port Severn, who provided a reliable source of labour for the company, was an even greater concern. Several French-Canadian bush workers who had migrated to Georgian Bay from Quebec and who were employed by Georgian Bay Consolidated Lumber had taken wives and "squatted" on the north side of the Severn River in Baxter township, in what was then the unorganized District of Muskoka. They had cleared small plots of land, built houses and stables, and generally lived, they thought, in relative security. The men worked in lumber camps in winter, on log drives in spring, and in and around the mill in summer.

In 1878, the provincial government, with a view to selling land in the north, had Baxter and other townships in Muskoka surveyed and laid out in lots. The security of the French-Canadian squatters was then threatened. At worst, the land on which they lived, but to which they did not hold title, could have been sold out from under them; at best, they would have had to buy their land, and this they could not afford to do.

In response to a petition from Georgian Bay Consolidated Lumber and others (similar situations existed in other townships in Muskoka and Parry Sound), the Ontario government passed an order-in-council in 1880 adding Baxter township (and several others) to the free grant lands set aside under the Free Grants and Homestead Act of 1869. The French-Canadian squatters were then eligible to apply for patents to their land, provided they met the requirements for settlement laid down by the act.

Before 1878, twenty-six families were residing in the Baxter township suburb of Port Severn; by 1883, the number had increased to forty families. Confirming the locations of the original squatters and the new legitimate settlers was not an easy task. In some cases the land the settlers had cleared and now claimed bore little relationship to the lots drawn up by the survey of 1878, on the basis of which the patents were to be granted. Some clearings were on more than one lot; some settlers claimed the same land, resulting in much conflict and hard feeling among them.

By 1885, because no patents had been issued, James Scott wrote to Ontario's commissioner of crown lands asking that "a surveyor or some person competent to follow the lines on the ground"[32] be sent to Port Severn to sort out the boundaries so that the claims would be correct and fair to all. A surveyor was sent to the community at the end of September; boundaries were marked out, affidavits taken from settlers, and patents subsequently issued.

Port Severn was not as large as Waubaushene, nor did it have the sort of recreational and cultural facilities of the larger centre. About one hundred families lived at Port Severn in the hey-day of the lumber trade. The heart of the village was on the company's property, on the south side of the river, where the principal buildings were located — store, warehouses, office, boarding-house, managers' houses, churches, and school. No direct road connected the two villages, because the company would not allow construction of a bridge across Matchedash Bay that would interfere with the movement of tugs and barges in front of the mill and piling yard. There was a good deal of communication between the two villages, however — by boat in summer and by winter road across the ice, and after 1887 by telephone. Despite its isolation, Port Severn was a prosperous and vital community.

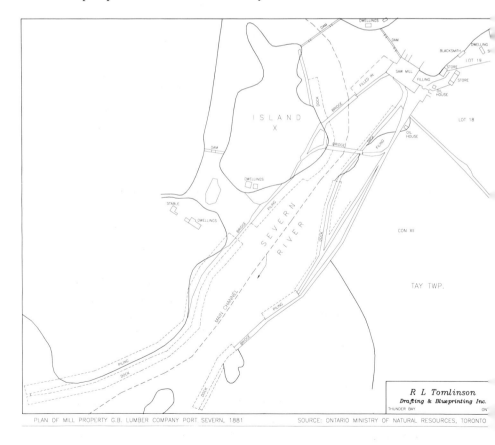

PLAN OF MILL PROPERTY G.B. LUMBER COMPANY PORT SEVERN, 1881   SOURCE: ONTARIO MINISTRY OF NATURAL RESOURCES, TORONTO

Chapter Twelve

Arthur's Team

Arthur Dodge was generous, likeable, even popular with his friends and employees, but he lacked the shrewd business sense of his father and older brother William. A spendthrift, like Anson, he lived well beyond his considerable means. Finding it necessary to borrow money to pay for his and Josephine's expensive tastes, he was constantly in debt. Not only did he have the two large houses at Waubaushene to maintain (three, after the Present Island cottage was built), but he had the rented mansion on East 34th Street in New York to keep up, and, in 1890, he built yet another grand summer mansion, in Weatogue, near Simsbury in Hartford County, Connecticut. His sons were sent to expensive private schools.

Left to his own devices, Arthur would probably have either bankrupted Georgian Bay Consolidated Lumber or lost it to creditors. But thanks to the close supervision of his activities by his director-brothers and the wise counsel of his knowledgeable and practical-minded Canadian managers, which Arthur had the good sense to follow, the company prospered.

There were minor fluctuations in the lumber market, from time to time, and these affected earnings, but, generally, the market remained strong through the 1880s, as the commissioner of crown lands had predicted in 1879. Overall profits of the company were substantial, providing good incomes to shareholders, while leaving enough to retire the $500,000 bond issue in a little over five years; in 1892, William Earl's executors were able to discharge the mortgage and trust deed given by the company as security for the bonds.

But despite the firm's profitability, the directors were determined to sell it, in order to redeem the $371,000 in stock held by

William Earl's still-unsettled estate. The proposed sale in 1891 of the Page mill and Wanapitei timber berth to Lynn and Hammond and Sibley and Barings, respectively, had fallen through, but in 1892 another Michigan firm, Merrill, Ring and Co., of Saginaw, offered to buy the entire company. The selling price was $1.5 million — $1 million in capital stock and $500,000 for the mills, timber limits, and other properties.

A sale would be complicated, because Georgian Bay Consolidated Lumber had a lien on A.M. Dodge and Co. of Tonawanda for a $200,000 indebtedness incurred by Arthur when the firm was established. As early as 1890, the directors had ordered Arthur to liquidate the debt, which he had been attempting to do with as much of his one-third share of the annual dividends as he could spare and from monthly lumber sales of A.M. Dodge and Co. But there was still a substantial amount owing to the company when an offer to purchase was made. Another complication was that Traders Bank held fifty shares in Georgian Bay Consolidated Lumber — stock that Arthur had put up as collateral on a $50,000 personal loan.

In 1891, Merrill and Ring bought the Wanapitei timber berth, with its estimated 500 million feet of standing pine, and it also bought the previous winter's harvest of 19 million board feet of saw logs, logs intended to be cut in the Anson mill but then floating in the inlet in front of the mill's charred remains. The exact purchase price of the timber is not known but was probably about $300,000. It had been Merrill and Ring's original intention to boom the logs across Lake Huron to its mills at Saginaw, as many Michigan companies were doing, but because the American tariff on Canadian lumber had been reduced to one dollar per thousand feet, and later dropped altogether, the company decided that it would be more profitable to saw the logs in Canada and export the lumber than to risk the hazards of towing the logs to Michigan. Consequently, in 1892, it bought all the Maganettewan property — the Page mill, the vacant Anson mill property, and the remaining timber berths — for $450,000 and took out an option to buy the three other mills. Probably because of financial difficulties encountered during the stringent market conditions of the mid-1890s, Merrill and Ring did not take advantage of the option.[1]

Arthur then decided to buy the three remaining mills — Port Severn, Waubaushene, and Collingwood; with these he planned to incorporate yet another Georgian Bay Lumber Co. Acting at arm's length, he bought the mills, premises, and associated timber limits from Georgian Bay Consolidated Lumber for the very good price of $150,000. His New Jersey manager, William J. Hunt (Anson's former partner), actually bought the property and signed it over to Arthur the same day — 4 June 1893. With over $1 million in cash in their possession, but no property, the directors were able to liquidate the company. James Scott was appointed liquidator on 6 July.

Because the liquidation and closing down the company were carried out as a "voluntary winding up" no application to the court was necessary, so no public records of the proceedings are available. It is clear, however, from such records as do exist, that with the cash realized from the sales, with the approximately $125,000 in insurance from the Anson mill and the Waubaushene box factory, and the surplus in the profit and loss account, each shareholder received full payment for his shares as well as his portion of the outstanding dividend notes.

Arthur came out of the transaction with little or no cash for his 3,520 shares, but he did own the three Georgian Bay mills and held clear title of the Tonawanda property, and the $50,000 debt with Traders Bank was liquidated. Arthur's untimely death in 1896 delayed the winding up, so that it was not until March 1901 that James Scott was able to advise the provincial secretary that the winding up had been completed and all assets distributed among the shareholders.[2] The company's charter was then surrendered. Meanwhile, Georgian Bay Lumber Co. Ltd. had been incorporated.

On 21 April 1893, an application was made to the Dominion secretary of state for letters patent incorporating The Georgian Bay Lumber Company (Limited) with capital stock of $200,000, divided into 2,000 shares of $100 each. Letters patent were issued on 13 May, incorporating the company and empowering it to purchase the mills, premises, timber limits, and all the rights and privileges of Georgian Bay Consolidated Lumber, and authorizing it to carry on the lumber trade in all its facets. It was also empowered to purchase the business of the Dominion Mercantile Co.

Two features of the incorporation are significant. First, the company was incorporated under Dominion rather than provincial charter, as all the Dodges' earlier Canadian companies had been. This was done probably because Georgian Bay Consolidated Lumber had not even started the process of liquidation; its charter was still active. To avoid confusion, a Dominion incorporation was sought. Further, because the new company would, like its predecessor, carry on most of its trade in the United States, it made sense to hold a charter under Dominion law.

Second, and the most significant feature, the charter assigned little of the share capital to Arthur. Because he personally owned all the mills, timber limits, and equipment around which the new company would be formed, Arthur could have acquired, through transfer of the mills, the bulk of the shares, say 95 to 98 per cent, allowing each of the Canadian managers to hold a few shares, thus "tying them to him" and ensuring "special zest" in their work. This was the kind of arrangement that the Dodges had made with senior management in the past.

But Arthur was, as usual, short of money. So when he transferred the mills to the company, he took only 1,075 shares — about 53 per cent — of the stock; for the balance he took cash. Under an agreement between Arthur and Georgian Bay Lumber, dated 6 June 1893 (two days after Arthur acquired ownership of the mills), he sold the mills to the company for $200,000. The company paid him $103,250 in cash, and for the $96,750 balance, together with the $10,750 that he had paid in at the time of incorporation, the company issued him 1,075 fully paid-up shares. The balance of the stock was subscribed for by seven Canadians: Henry L. Lovering, W.J. Sheppard, and James Scott each took 225 shares; Thomas Sheppard took 110 shares; W.H.F. Russell, Charles P. Stocking and Joseph W. Hartman subscribed for 75, 40, and 25, respectively. Arthur was president, James Scott treasurer and vice-president, Charles Stocking secretary, and W.J. Sheppard became general manager. Arthur, Russell, Scott, W.J. Sheppard, and Stocking were directors.

No longer were the Canadian managers merely token shareholders and nominal directors; they were full-fledged, influential members, with a significant voice in determining the future direction of the company. Perhaps no shrewder or more competent group of lumbermen, accountants, and storekeepers ever took over management of a lumber company. Under its capable administration Georgian Bay Lumber would eventually prosper as few other firms would, giving it a reputation, all over Canada, as one of the finest lumber companies in Ontario. The Waubaushene mill would be "known for many years as one of the largest and best equipped"[3] in the Georgian Bay district. Whether the company would have fared as well had Arthur maintained outright ownership is arguable, but two facts are certain: by giving up 47 per cent of the company to the aggressive Canadians, Arthur ensured the company's continued success; in keeping only 53 per cent for himself, he allowed his widow and his family to become merely rich, and not wealthy, from the considerable future earnings of this most profitable company.

To return to 1893, the company had some difficulty in taking over the business of the Dominion Mercantile Co., which, it will be recalled, had since 1888 operated the stores on which the company relied for stocking its bush camps and feeding and clothing the villagers. The Dominion secretary of state was reluctant to transfer to a lumber company the wide powers to engage in business as general merchants throughout Canada, formerly possessed (but not used) by the Dominion Mercantile Co.

The secretary of state wanted the company's authority limited to supplying goods, wares, and merchandise to meet the requirements of its employees only. The company objected to the limitation, which would prevent it from providing goods to saw-log jobbers, buying produce from settlers, selling goods to them in return, and generally doing business with "others in the districts with whose development [the company] was so closely identified."[4]

In the end, the secretary of state compromised. The letters patent, when issued, permitted Georgian Bay Lumber to purchase the business of the Dominion Mercantile Co. "provided, however, that the operation . . . be confined and restricted to the districts within Canada wherein the mills of the Company are or may be for the time being situated."[5] That now meant Waubaushene and Port Severn.

The new company was formed in the depth of another economic depression which had started in 1891 and continued for several years. Consequently, the directors immediately set about consolidating the company's resources and disposing of unproductive real estate. In September a public auction was held to dispose of some of the timber lands the company held either under licence to the Ontario government or in fee simple. While the lands sold still contained a small amount of timber, the previous Georgian Bay Consolidated Lumber Co. had not harvested them for years but had been paying ground rent or taxes on them. Ten timber berths in the townships of Morrison, Muskoka, Ryde, Vestra, and Wood in Muskoka district and 1,838 acres of land in the townships of Orillia, Sunnidale, Tay, and Tiny in Simcoe County were put on the auction block.

Next, the company sold the Collingwood mill to the firm of Toner and Gregory. Georgian Bay Consolidated Lumber had not operated the mill for years, preferring to rent it to former mill manager D.E. Cooper and his partner, Melville, to whom the company also sold sawlogs from the Moon River timber limits. Because the mill and equipment were in poor state, the sale netted the company only $9,000.

The depression hurt the Canadian lumber trade generally, and the new Georgian Bay Lumber Co. suffered along with all the other firms. There was, however, one positive outcome of the depression that somewhat eased the lot of Canadian producers. A momentary supremacy of free trade ideas in the United States resulted in the removal of all duty on Canadian lumber. Earlier, Canada had revoked an export tax of two dollars per thousand board feet on sawlogs; thus, between 1893 and 1897, there was free trade in lumber and sawlogs between the two countries.

Under normal conditions this would have produced a marked expansion in Canadian lumber exports, but because of the depression the free-trade arrangement had little impact, as the demand for lumber dropped sharply in both countries, and prices fell accordingly. The combined maximum cutting capacity of the Port Severn and Waubaushene mills was about 50 million feet per year (Port Severn, 15-20 million feet; Waubaushene, 20-30 million feet), but rarely did the mills operate at anywhere near full capacity. Thirty million feet was the normal average, but during the depression they cut considerably less.

Removal of tariff barriers between the two countries harmed Ontario's lumber industry in another way, particularly in reference to the rapidly diminishing supply of pine. Since the early 1880s, Michigan producers had been buying up timber rights in Ontario. By 1886, Michigan men were reputed to hold 1.75 billion feet of standing timber in the Georgian Bay district,[6] much of which, in due course, would be exported as unmanufactured sawlogs. The export duty of two dollars per thousand feet on logs imposed by the Canadian government (increased briefly to $3 in 1888), generated large revenues for the Canadian treasury but did little to check the drain on Ontario's resources.

In truth, the Ontario government seems not to have been unduly alarmed by the magnitude of American ownership of natural resources: income from timber dues was substantial, and American bidders tended to increase the level of the bonuses paid for timber limits at public auctions. Canadian lumbermen, however, quite naturally resented the competition from those who could take home raw material to be manufactured behind a tariff wall and sold against Canadian lumber in a protected market. More serious, Canada was exporting jobs to the United States, although few people seemed to consider that a problem then.

Reduction of the American tariff to one dollar per thousand board feet on Canadian lumber in 1890, followed by removal of all tariffs in 1893, put the producers on a "level playing-field," so far as competition for markets was concerned, and encouraged some Michigan producers, like Merrill and Ring and the Emery Lumber Co., to remove sawing operations to Canada. But large-scale export of sawlogs continued. In 1892, after Canada had removed the export tax, it was estimated that 160 million feet of Ontario sawlogs had been taken out of the Georgian Bay area by producers in Saginaw and Bay City, Michigan. The concomitant effects of the depression, forcing some small mills on Georgian Bay to close down and larger ones to cut production, led to a hue and cry for reimposition of the export tax, even though the lumbermen's troubles were caused more by the general state of the times than by the export of logs.

The Canadian government seems not to have been concerned about what amounted to a raid on Canadian forests by American lumbermen, as a unique piece of legislation, passed in 1894, suggests. A group of nine Canadian mill owners on Georgian Bay, including James Scott and W.J. Sheppard, and six Michigan producers petitioned for incorporation of the French River Boom Co. Ltd. The new firm would build slides and booms on the Wanapitei and French rivers to handle sawlogs being run down the rivers to the owners'tow-booms in Georgian Bay. The act of incorporation, which also empowered the company to levy and collect tolls, received royal assent on 23 July.

The capital stock of the new outfit was set at $50,000, divided into one thousand shares of $50 each, to be allotted between Canadian and American shareholders in the ratio of 600 shares to Canadians and 400 to Americans; at no time could more than two-fifths of the stock be held by aliens or persons resident outside Canada. The provisional board of directors numbered seven, of whom four were required to be Canadian and three American. If, in the future, the board were reduced to five, three directors had to be Canadian.

This act was unusual in several respects. First, the French River Boom Co. was incorporated under Dominion and not provincial law, even though the issue of jurisdiction over driving rights on rivers and streams had been resolved in the province's favour in 1884, when the Privy Council upheld the validity of Ontario's controversial Rivers and Streams Bill of 1881. Indeed, the driving and boom companies on the Severn, Muskoka. and Magnetawan rivers had all received provincial charters. It seems, however, that the Dominion government, which had constitutional control over navigable waterways, claimed jurisdiction over the historic Lake Nipissing — French River voyageur route, giving it power to grant a charter to the new company. There may have been some basis for the claim: the Dominion-chartered Georgian Bay Ship Canal Co. was drawing up plans to canalize the route. In any event, the Ontario government did not challenge the legislation.

Second, and of even greater significance, the French River Boom Co. was incorporated by special act of Parliament, not through the normal process of letters patent issued by the Department of the Secretary of State. The Canada Joint Stock Companies Clauses Act, under which authority the letters patent would have been issued, did not provide for control of the distribution of capital stock by the nationality of shareholders. The Canadian government appears to have been determined, despite its willingness to allow American lumber companies to export large quantities of sawlogs, to prevent ownership and control of such a vital facility as timber slides and booms on a national waterway from falling into the hands of Americans; hence the special act. The act did waive a provision in the Companies Clauses Act requiring directors of companies to own stock absolutely and in their own right. In effect, lumber companies could hold stock in the French River Boom Co. and their representatives could act as directors. In fact, the Georgian Bay Lumber Co.subscribed for stock, and W.J. Sheppard was a director.

The new directors of Georgian Bay Lumber had more pressing matters than cat-and-mouse tariff competition between the American and Canadian governments; the tariff war did not much affect their business, at least not yet. Of more serious concern was loss of the Port Severn mill. During a hailstorm on the morning of 17 August 1896, the mill house was struck by lightning, igniting a fire that burned the mill to the water level, taking with it a tug, the store, and a warehouse. Only the strenuous efforts of the workers prevented the fire from spreading into the lumber yard; thus several million board feet of lumber was saved. But for the heavy rain that followed the hail, the whole village might have been swept away in the blaze.

While the unexpected destruction of the mill was a serious blow to the one hundred families at Port Severn whose livelihoods depended on it — especially to the French-Canadian settlers — it was not as great a loss to the company as one might suppose. The firm's timber limits on the Severn River then contained only enough wood to supply the mill for another two years at most; after the timber was gone, the mill would have had to be sold or closed down. Certainly the company would not have been able to continue operating it without incurring great expense.

At Port Severn, logs were fed into the saws from a log pond in the river on the upper level. To supply the mill from log booms brought in from the Georgian Bay side would have required portaging the logs up into the river or turning all the equipment around and relocating the jack-ladders to the downstream side. Because the piling yards, loading ramps, and wharfs were built on the shore, at the entrance to the river, these facilities also would have had to have been relocated, to make way for the supply booms. Prospective buyers would have faced the same problem, so that the mill's selling price would not have been very high. The $50,000 that the company received in insurance payments was probably more than could have been realized from a sale.

Under the circumstances, there was no possibility of the mill being rebuilt. The few remaining logs from the Severn River timber limits were towed to the mill at Waubaushene. And so, after the Port Severn fire, Anson Dodge's once-extensive lumbering network of nine mills, producing 80 million board feet of lumber per annum, was now reduced to one mill with a cutting capacity of only 24 million feet per year.

After the market for lumber improved, the company replaced the Port Severn mill, in 1901, with a smaller one, built in front of the main mill at Waubaushene. Known as "the small mill" because of its dimensions, it was used for sawing large-diameter logs, which were becoming scarcer around Georgian Bay; small-and medium-sized logs were cut in the main mill. The cutting capacity of the two mills was about 35 million feet.

After the fire at Port Severn, the English-speaking workers — manager, foremen, filers, mechanics, clerks, and storekeepers — were transferred to Waubaushene. Several of the houses were moved to Waubaushene with them. The remaining houses were sold to the Chew Lumber Co. of Midland and transported to that village. The Methodist church was moved to Fesserton. Only the Catholic church, the school, and the office building, later converted into a store, were left. Most of the French-Canadian settlers, with no place to go, stayed where they were.

Economically, the French suffered most from the loss of the mill. The young men continued to work in the winter lumber camps and on the spring log drives, but there was little employment for them in the village in summer, until after the Trent Canal lock was built in 1915, stimulating a brisk tourist trade. Then many residents found work as tourist guides and as summer help in the several lodges that sprang up near the village.

A greater blow to the directors than the loss of the Port Severn mill was the unexpected death of Arthur Dodge. Like his father, Arthur had suffered from gout and was frequently bedridden with it, but it was always believed that he enjoyed good general health; certainly his active social life led his friends to believe that.

In February 1896 Arthur had become ill from what was later diagnosed as dropsy. After spending some time in a New York hospital, he had recovered sufficiently to take his family to Present Island in the first week of July. Later, not feeling well, he returned alone to New York for medical treatment. His condition deteriorated and Josephine and Marshall were summoned to New York. Again Arthur's health improved; he was allowed to return to Weatogue, where it was thought his health was being restored. He took a turn for the worse, however, and died suddenly on 17 October. He was only forty-four.

Believing that Arthur was going to recover, Josephine was totally unprepared for his death. It was said that her hair turned white within a week from grief and shock. An even greater shock awaited her when she discovered the state of Arthur's finances. The value of his assets, appraised shortly after his death, amounted to $561,853.04, but his debts amounted to $294,881.77, leaving a net value to the estate of $266,971.27. Josephine was faced with a dilemma. By the terms of Arthur's will, she was to receive the house at Weatogue and one-quarter of the balance of the estate; a second quarter was to be placed in trust, the income from which was to be used for her support during her lifetime, after which the capital was to be divided among their five sons. The remaining half of the estate was to be placed in a trust until Marshall reached the age of twenty-five, when it was to be divided equally among the children. The bulk of the estate was tied up in various stocks; some, like those of Georgian Bay Lumber, paid good dividends. Had the executors, of whom Josephine was one, been forced to sell the stocks to pay the debts, there would have been little of the estate left to distribute when Marshall turned twenty-five, and very little income generated in the mean time.

Josephine met with the major creditors, many of whom were members of Arthur's family and close friends (he owed W.J. Sheppard $35,000) and asked for a period of time within which to pay back the debts. This given, she set about the task of administering Arthur's estate, herself, from a small, rented office on Wall Street. She managed extremely well, impressing her male associates with her shrewdness and keen business judgment and proving to her in-laws that she was made of tougher stuff than they had ever imagined.

To cut expenses, Josephine moved out of the large rented mansion on East 34th Street, first into the Abelmare Hotel on West 23rd Street and then into a duplex on Park Avenue. She sold the Present Island cottage for $8,000 to two sisters from Toronto; *Skylark* was sold for $3,000 and *Ida* for $800. She sold many of Arthur's personal possessions — his horse, hansom, and bicycle and the contents of his wine-cellars — and she disposed of his bonds and shares in riding, steeple-chase, and hunt clubs. Some unprofitable stocks were sold, and others reorganized. The probate judge allowed her an allowance of $10,000 per year from the estate's income to support her and her family.

By 1901, when the estate was to have been finally distributed (Marshall had turned twenty-five), Josephine had paid all the debts but $45,000. Even these would have been cleared had the dividends from Georgian Bay Lumber not been greatly reduced in 1899 and 1900 because of expensive purchases of timber limits on the north shore of Lake Huron — acquisitions that Josephine had approved because of their future value. The probate judge granted an extension of twelve months for settling the estate; the remaining creditors (D.Stuart Dodge, her son Marshall, and a close friend) had not objected. Two more extensions were later granted.

Finally, in 1904, Josephine was able to make a partial distribution of $188,293.89 worth of stocks. Distribution of the remainder of the estate, which was largely in the form of property, was held up because of complications encountered in the conveyance laws of New York state. By then (1904) all the debts had been settled, and the estate had a net value of $350,771.24, much of it in sound, dividend-paying stock.

Arthur's 1,075 shares of Georgian Bay Lumber were eventually divided among members of the family, with Josephine keeping 381 shares; Arthur's brother Stuart received 263 shares, in partial payment of a $100,000 debt (representing Stuart's stock in the former Georgian Bay Consolidated Lumber Co.); Marshall 156; Murray, Douglas, and Geoffrey, 75 each; and Percival, 50. Marshall and Murray became directors of the company, and Marshall was made vice-president. The Dodge family kept all its stock until 1925, when there was no more income and the stock was devalued to one dollar per share. In the intervening years, it collected hundreds of thousands of dollars in dividends.

Josephine never returned to Waubaushene after Arthur's death. Marshall went back occasionally to attend board meetings, particularly if major decisions were to made. His last known visit was in 1921.

In 1913 the Alexandra House burned down. Josephine then sold her Waubaushene home and all the other small lots to the company for $3,000. Dodge House became the company's premier boarding establishment, managed by Mrs A.L. Bettes, who had run Alexandra House since the 1890s.

During the years (1897-1905) that Josephine administered Arthur's estate and took it out of debt, the principal source of income had been dividends from Georgian Bay Lumber. In the six years between 1897 and 1903, the estate received $268,004.36 in dividends on only 837 shares. (The other 238 of Arthur's original 1,075 shares had been signed over to Josephine before Arthur's death. Income from those is not included in the figure quoted above.) The dividend income averaged slightly more than $53 per share per year, or a 53 per cent average annual dividend. That was exceptional profit-making, considering that the period saw a slump in the lumber market and that the company had lost the production of the Port Severn mill and had spent considerable revenue on timber limits and construction of a new mill at Waubaushene.

The man chiefly responsible for the company's phenomenal success was W.J. Sheppard, general manager at the time of Arthur's death, and later president.

Part Three

Third Empire: The Sheppard Family (1896-1942)

William J. Sheppard 1854-1934   – courtesy Gladys Kelly

Chapter Thirteen

W.J. Sheppard and His Clan

It was important for Josephine Dodge to select some competent person to take charge of the Georgian Bay Lumber Co. after Arthur's death in 1896. She was not capable of running the company herself, and twenty-year-old Marshall, with no business experience or training in the lumber trade, was not qualified to succeed his father.

Of the three senior shareholders — Lovering, Scott, and Sheppard — Vice-President Scott was the most obvious successor. Scott had been secretary treasurer and a director of the Dodge lumber companies for many years. He was at home in the business and financial world, was familiar with corporate law, and had plenty of connections in both business and politics; indeed, he had been pro- posed as a Conservative candidate in the provincial election of 1890.

Although he had no practical experience at the operational level (none of the Dodges had any bush experience either), Scott understood the lumber business as few other men in Ontario did. Moreover, he had been extremely loyal to the Dodges and had their unqualified trust. But Scott was already fifty-eight years of age and living in semi-retirement in a new home in the Parkdale district of Toronto. Whether Josephine asked him to assume responsibility for running the company is not known, but if he was asked, he refused.

Sixty-two-year-old Henry L. Lovering, having spent his whole life in lumbering, was the most experienced lumberman of the three. He had worked for the Dodges since incorporation of the first Georgian Bay Lumber Co. and had been in charge of woods operations for twenty years. He had been a minor shareholder and director of Georgian Bay Consolidated Lumber and owned and operated his own lumber and shingle mill at Coldwater. However, his age would probably have ruled him out, if he had been interested in the position.

And so it was W.J. Sheppard whom Josephine summoned to New York to ask him to assume active command of the company for which he had worked since he was eighteen years old and through the ranks of which he had climbed — Horatio Alger style — from clerk to general manager. In choosing Sheppard, first as surrogate for Arthur, later as president in his own right, Josephine displayed the executive quality that permitted her, a bereaved widow with five children, to manage so successfully Arthur's estate in a tough, competitive, man's world — the ability to recognize and select good men for positions of responsibility. Although Sheppard would never hold more than 22.5 per cent of the stock while the company was productive, he always managed the firm with as much diligence and close attention to detail as if it had been exclusively his. Over the course of the next twenty-five years his name would become synonymous with Georgian Bay Lumber; both he and the company would be regarded as symbols of lumbering success in Ontario. Indeed, many contemporaries would have been surprised to learn how little of the company was actually his.

Sheppard's life, from poor farm boy to millionaire, was a Horatio Alger story come alive. He followed the same mythic formula possessed by Alger's young heroes whose exploits delighted millions of nineteenth-century readers: luck, pluck, decency, and hard-work. In assessing Sheppard's distinguished career, a reporter for the *Canada Lumberman* could have been writing about any one of dozens of Alger's fictional heroes: "The position which he [Sheppard] occupies to-day in the lumber arena has been won by pluck, integrity and perseverance, as from a very early age he had to make his own way in the world. This he did by hard work and keeping everlastingly at it."[1] And Sheppard's own explanation for his achievements, given in another context, could have been straight from Alger: "When I was given any business, I looked after it. I went wherever I was sent, and took the job they gave me. I never asked for a position but just took them as they came."[2] His approach to business success was vintage Alger: "Those that start at the top generally get to the bottom, and those that start at the bottom, if they stick to their business, have a good chance of getting to the top."[3]

William Joseph Sheppard was born in 1852 on a farm near Keswick, Ontario, son of Irish immigrant Richard Sheppard and his Canadian-born wife, Miranda Sprague. "Red Billy" friends nicknamed the precocious young Irishman with the red hair and determined ways, who after receiving a rudimentary education in the village school, set out at fourteen to earn his own living. His first job, which he held for four years, was driving a stage-coach between Beaverton and Newmarket. A frequent passenger on his stage was A.G.P. Dodge.

In 1870, his uncle Daniel Sprague, minor shareholder and book-keeper-treasurer of the first Georgian Bay Lumber Co., gave Red Billy a clerical job in the Orillia office, where his older brother, Tom, had already been engaged. Red Billy proved bright, a fast learner, and good with numbers. Manager W.J. Macaulay soon recognized his abilities and gave him his first practical experience in the rough and tumble of a lumber shanty; it was in Medonte township, between Orillia and Coldwater, just off the Coldwater Road.

His job was that of a scaler (or culler, to use the nineteenth-century term), an important member of the shanty team. He would count and grade the logs that were cut and skidded each day, measure the diameter of each, and, using standard tables, convert its volume into board feet equivalence. Scalers' records determined the timber dues to be paid to the Ontario government. Provincially appointed woods rangers, or inspectors, visited lumber camps periodically to verify the accuracy and honesty of scalers' reports. The scaler also kept a record of the men's time and a daily inventory of camp supplies.

Company managers were so impressed with Red Billy's competence in record keeping that at the end of the logging season they sent him to Waubaushene to work in the store. After another winter in the bush he was sent to Port Severn to manage the store, keep accounts, and make out bills of lading and lumber invoices. For the next few years, his duties alternated between scaling in winter bush camps and doing clerical work at Port Severn in the summer.

In 1874, he married Ellen Mary Buchanan, daughter of Robert Buchanan of Coldwater and granddaughter of pioneer contractor Jacob Gill. The Sheppards lived at Port Severn for the next few years, as W.J. gained an expanding role in company matters. (William Earl Dodge had taken over the presidency of Georgian Bay Lumber.) "I never had a happier time than at Port Severn when there by ourselves. I used to work eighteen hours a day,"[4] Sheppard recalled wistfully, in a rare personal interview with the press on the eve of his fiftieth wedding anniversary, in 1924.

In the late 1870s, he was sent to Gravenhurst, in charge of the company's operations in Muskoka. While he was there his oldest children were born. Because Ellen Sheppard did not favour raising a family in an isolated logging community, especially with her husband absent in the bush for long periods, Sheppard decided to establish a permanent home in Coldwater, where Ellen could live near relatives and friends.

For the rest of his lumbering career, with considerable sacrifice of the pleasures of normal family life, Sheppard spent many months of each year away from home, attending to the various jobs assigned him by his employer in its far-flung operations: overseeing winter logging operations, supervising summer river drives, managing a mill here or stores operation there.

In 1888, when only thirty-six years of age, he was given his first major management position, when Arthur Dodge sent him to isolated Byng Inlet, for Georgian Bay Consolidated Lumber, to superintend the recently reactivated Page mill and the underworked Anson mill. Under Sheppard's expert superintendence, these hitherto unprofitable mills increased outputs to become two of the leading producers in the company's chain of sawmills, becoming highly attractive facilities to prospective buyers. Then, recognizing that Sheppard had operational management skills, superior to his own, Arthur promoted him to general manager of the whole operation in 1890. Sheppard moved to Waubaushene, where he lived in the company's boarding-house during the week, travelling by train five miles to his home in Coldwater each weekend — a routine he followed for the next forty years. He was general manager of Georgian Bay Consolidated Lumber for the three remaining years of its existence and assumed the same position with Georgian Bay Lumber Co. Ltd, when it was formed in 1893, with only two mills under his jurisdiction.

The exact date on which Sheppard officially became president of Georgian Bay Lumber is not known. The secretary of state did not, prior to 1902, require Dominion-incorporated jointstock companies to submit annual returns listing officers and directors. Such returns as may have been submitted by Georgian Bay Lumber between 1902 and 1919 have been lost; therefore no official record of Sheppard's promotion to president exists. One account[5] gives 1907, while another[6] places the date at 1905; 1904 is probably more accurate.

Sheppard bought James Scott's 225 shares when he died in 1902.[7] In 1904, Josephine distributed the stock held by Arthur's estate among several members of her family, leaving Sheppard, with 450 shares, the principal shareholder. With Josephine's backing, Sheppard had no trouble securing enough votes to obtain the presidency. Also about 1904, Marshall Dodge became director and vice-president, Murray a director, and Charles Stocking both treasurer and secretary.

If Sheppard had a role model for his ascent in the perilous realm of industrial capitalism, it was probably William Earl Dodge. He probably never met William Earl, not being stationed high enough in the company hierarchy to warrant an introduction during any of the president's infrequent visits to Waubaushene, but Sheppard certainly knew the elder Dodge by reputation, and he would have met Melissa Dodge when she spent her summers at Waubaushene with Arthur. He had seen the amazing results of William Earl's business genius as he shored up Anson's crumbling corporate edifice — the son being one of the unfortunates whom Sheppard had known "to start at the top and get to the bottom."

Sheppard kept the oil painting of William Earl that Arthur had placed in the president's office many years before. Later, a photograph of Ellen Sheppard's prune-faced grandmother, Sarah Parmer Sutherland Gill, was added, by her greatgrandson, William Gill, who became manager of the Waubaushene mill when J.C. Else retired. Whether sheppard found inspiration in the stern countenances or saw no reason to change the office decor, or perhaps because he noticed that the austere faces made visitors uncomfortable and hastened their departure, the pictures remained there all through Sheppard's presidency and for many years after he was dead.

Like many businessmen, Sheppard was a pacer. A loner with no confidant to share the relentless pressure of business; a sporadic church-goer, with no spiritual refuge; a compulsive worker, with no hobbies; a teetotaller, who had never relaxed with a drink; and an introvert, with no social diversions, Sheppard walked off his frustrations.

A grandson remembered him pacing up and down the Waubaushene dock, chewing vigorously on a wad of tobacco and spitting nervously into the water, as he anxiously awaited the arrival of the steam tug with the log booms coming down from the north. The logs always came (not once in fifty years was the mill idle for want of logs), but there was always the possibility that a storm or a late break-up might delay the tug: the rapidly dimishing stock in the log pond would be consumed before the new logs arrived,leaving 400 mill hands on the payroll standing idle. And so, every spring, Sheppard paced the dock.

A granddaughter recalled that on rainy Sundays, when fishing was out of the question, her grandfather would pace back and forth on the veranda of his summer cottage, deep in thought. The grandchildren wondered: "What scheme is Grandfather cooking up, and who is he going to skin on Monday?" But Grandfather never "skinned" anyone on Monday, or Tuesday, or any other day of the week, for Sheppard was an honest businessman.

To be more precise, he worked within the parameters of the law, for business honesty to him was more an exercise in legal nicety than a moral imperative. He rarely did anything without keeping a detailed record of the transaction. He never gave anything away, he sold it. "Yes, you can build your boat-house on company property" (after the mill was gone), or " Yes, you can have that object" (if the object was worthless), "but let's make it legal. Give me a dollar," he would say to a supplicant. The dollar proffered, Sheppard would order his secretary to draw up an indenture that would be duly signed, witnessed, and sealed.

To Sheppard a dollar was more than a medium of exchange or even a shield against want; it was a talisman, a fetish to which he was excessively, almost irrationally, devoted. Perhaps that is why he accumulated so many of them and parted with so few. But on at least one occasion he voluntarily parted with some. At Christmas 1925, after the Waubaushene mill had been torn down, after his own lumber companies had been woundup, when his considerable investments carried a book value of millions of dollars, and with the joyous festivities of his fiftieth wedding-anniversary celebrations still cheering his heart, he was moved to perform a rare act of beneficence. He drew a number of crisp, uncirculated one-dollar bills from the bank and distributed them, as Christmas presents, among his loyal employees. Ella Breech, a spinster who had been his secretary-manager for twenty-five years, received one, with a hand-written note attached: "To Miss Breech — A penny for good luck. W.J. Sheppard." An astonished Miss Breech carefully folded her good-luck token and stored it and the note in a safe place. Today, her niece treasures the still-uncirculated dollar bill and the note.

Ladies having afternoon tea, Waubaushene 1896
– courtesy Margaret Bell

Tug Waubaushene *at Port Severn c1890*
– courtesy Pauline Webel

Unlike his model, William Earl Dodge, Sheppard did not tithe, or even contribute regularly to the church. Nor did he, like many of the robber barons of the late nineteenth-century, discover charity late in life and give generously to ease a guilty conscience. Because every cent he owned was earned honestly, through hard work, he felt no guilt in achievement and therefore no necessity to purge a soiled soul with liberal giving. He always lived unostentatiously, managed his own investments, and frugally watched the bills. Those who knew Sheppard described him as "tight"; they would have agreed that he was a confirmation of Hemingway's observation: "The rich are different from you and me; they spend less money."

Although the Georgian Bay Lumber Co. made small donations, annually, to the Waubaushene church and some organized charities (see Appendix H), it is doubtful if Sheppard, personally, ever gave more than token amounts. Only once did he make a significant contribution to a church, but that was prompted more by pride than philanthropy. Mrs Sheppard was a devoted member of the Methodist church; indeed, her mother, Rachel Gill Buchanan, was one of the founders of Methodism in Coldwater (1866).

A new church(the third Methodist place of worship in Coldwater), completed in 1911, was barely paid for when, in March 1923, fire completely destroyed it. The Methodist congregation then worshipped in the Presbyterian church, which was graciously volunteered by its members, while the Methodists' building committee pondered how enough money could be raised to replace their church. The thought of his wife relying for a place to worship on the well-meaning but narrow-minded Calvinists, whose ethos had dominated Georgian Bay Lumber for fifty years, was anathema to the independent-minded Sheppard. "You'll have your own church, even if I have to build one," he promised Ellen. True to his word, he contributed most of the $25,000 replacement cost. Within eight months a new church was built, fully paid for, and dedicated.

Sheppard may have made parsimony an art, but his stinginess was not unique; most lumber barons of his generation were imbued with like talents. In 1907, a public plea for building funds by the Hospital for Sick Children in Toronto was answered by the Georgian Bay Lumber Association with the princely contribution of $1,000 on behalf of its several members, even though the combined annual profits from the Georgian Bay lumber trade totalled millions of dollars.

Occasionally, Sheppard was capable of spontaneous outbursts of generosity, but even these acts, like the building of the church, had a question of"cui bono?" behind them. A derelict Indian, named Joe Cigarette, once approached him with a hand-made axe handle. Sheppard admired the workmanship. "I give you axe handle, you give me new clothes," Joe Cigarette proposed. Impressed by the uncharacteristic demonstration of entrepreneurship on the part of the Indian, Sheppard sent him off to the store with a chit for new clothing. But he kept the axe handle.

When the partially built Roman Catholic church and rectory burned in November 1914, Sheppard personally directed the removal of valuables from the chapel and provided a place to store them; he also provided stable space in the company's barn, free of charge, for the priests' horses, and he donated a vacant company house as temporary shelter. Some time later, he gave Father Bouvrette a second-hand car, and when he built a new boat he gave the priest the old one.

Like the gift of the mythical Greeks, Sheppard's presents to the Catholic clergy had a hidden purpose, or so it has been alleged. Sheppard was a staunch Conservative; the priests, and hence their flock, supported the Liberals. If he could convert the priests to the one true politic, he reasoned, he could garner a lot of votes for the Conservative cause in Simcoe East; hence the great expenditure of time and money on the Roman Catholic clergy of Waubaushene.

In the election campaign of 1911, Sheppard used cash in a more direct way to influence the outcome of the election. He desperately wanted the Conservatives under Robert Borden to oust the corrupt Liberal government of Wilfrid Laurier. Locally he wanted the Liberal incumbent, lumberman Manley Chew, to be replaced by William H. Bennett, the Georgian Bay Lumber Co.'s lawyer from Midland. The main issue was reciprocity, or free trade with the United States. Laurier had already signed such an agreement with the Americans, but the Conservatives opposed it, as they had "unrestricted reciprocity" in the election of 1891, fearing as they had then, loss of Canadian sovereignty and eventual absorption by the United States.

Although free trade would have been manifestly advantageous to the lumber trade — it always had been — Sheppard remained doggedly loyal to the Conservatives' position and sought ways to help them defeat Laurier's government. He placed posters of Borden and Bennett in the mill, the store, the office, and the logging camps, and just prior to the election he announced that the workers' wages would be increased from $28 to $35 per month if the Conservatives were elected. The French Catholic workers then faced a dilemma: to obey the church, vote Liberal, and keep their souls intact, or defy the priests, vote Conservative, and earn $7 a month more. They defied the priests. Borden, as we know, won the election. Bennett swept Simcoe East with a majority of 500 votes (women did not have the franchise then), the Georgian Bay workers got their $7 raise, and free trade became a dead issue for the next seventy-five years.

The frivolous art of popularity was not part of Sheppard's make-up. He was taciturn, reserved, penny-pinching — not qualities to win him close friends, but he was frankly in the business of making money. Such friends as he did have were business associates; there were plenty of those. His innate sense of duty and his commitment to law, order, and authority ensured his unswerving loyalty to Arthur Dodge during the years he served him, even though he may have privately disapproved of Arthur's profligate spending.

Sheppard remained both kind and deferential to Josephine, even after he had remade the Georgian Bay Lumber Co. over in his own image and after he had probably accumulated more wealth than any of the Dodges, except his one-time predecessor, Cleveland H. Dodge. He consulted Josephine on all major company decisions; he paid annual dividends promptly and explained the reasons for annual fluctuations in them. By making Josephine and her sons comfortably independent, he justified, many times over, the confidence she had placed in his ability.

In 1920, to mark his fifty years with Georgian Bay Lumber and as a token of her appreciation, Josephine presented him with a handsome, gold-encased fountain pen with the inscription "W.J. Sheppard 1870-1920." A congratulatory telegram, sent by Marshall Dodge on the occasion of his golden wedding anniversary, read in part: "My mother and all the family extend to you and Mrs. Sheppard our heartfelt congratulations. You are the only one left of my father's close friends, and the one friend to whom all the family owe so much."[8] It was a much-deserved accolade.

Sheppard may not have been universally loved, but all who knew him respected and admired his business acumen, even if they sometimes envied its prodigious fecundity. Many companies took advantage of his astute and discriminating investment sense by placing him on their boards of directors, sometimes making him a vice-president or even president. One of his first directorships was with Traders' Bank, with which Georgian Bay Lumber had done business since its inception. When Traders' Bank was taken over by the Royal Bank of Canada, Sheppard was made a director of the latter institution, an honour he maintained until his death. Small-fry investors, employees, and acquaintances with a little money to invest sought his advice, which he gave freely. His secretaries, Agnes Anderson and later Ella Breech, did quite well financially by placing small amounts in investments recommended by "the boss."

Of course, not all his "tips," especially those in penny mining stocks, paid off. But those who ignored his advice to invest in the McIntyre gold mine at Porcupine, Ontario, with whose development Sheppard had been connected since the beginning, regretted it. One stranger took his advice and was grateful. The person, a travelling salesman, happened to give Sheppard a ride from Waubaushene to Toronto one day. He was surprised when the slightly unkempt stranger in his car proceeded to convince him of the desirability of investing in what was then a little-known mining property in northern Ontario. The salesman was so impressed with the sincerity and apparent knowldge of his passenger that when he got home he sold his house and bought McIntyre stock; he became rich when the stock shot up in value just as the hitch-hiker had predicted it would.

Sheppard got his start in lumbering. The capital he acquired in that industry permitted him to move into the much more speculative, riskier, but potentially more profitable realm of financial capitalism. He proved as competent in the latter as in industrial capitalism; so, while lumbering made him rich, it was his financial wizardry that made him wealthy.

When he took over management of Georgian Bay Lumber, however, there was no assurance that it would make him rich, or even that the company would survive. He had barely moved to Arthur Dodge's desk when he was confronted with the possibility of a recurring difficulty — typically faced by industries, like the sawn lumber trade, depended on one external market — imposition of a nefarious American tariff. Free trade in wood products which had emerged as lumbermen on both sides of the border tried to cope with the depression of the mid-1890s was threatened. As often occurs in the United States during depressions, a strong protectionist movement began to grow, reaching a peak of intensity during the presidential election campaign of 1896.

After the election, Maine Congressman Nelson Dingley (1881-99), who chaired the House Ways and Means Committee, drafted a bill that would impose new tariffs on silks, linens, and chinaware, increase existing tariffs on hides and wool, and reimpose an 1890 duty on metals. Lumbermen, especially in Michigan, persuaded Dingley to add lumber to the list. Lumber producers in Michigan had become disturbed about the increasing volume of duty-free, and in many cases superior-quality, Canadian lumber entering their traditional market areas. Also, many in Michigan resented the movement of sawmills into Ontario, creating unemployment in Michigan mill towns.

The man most influential in persuading Dingley to include reimposition of the lumber tariff in his trade bill was Russell Alger, former governor of Michigan (1885-87), and secretary of war (1897-9) in William McKinley's cabinet. Alger was also a lumberman, operator of a sawmill at the outlet of the Black River, approximately twenty miles south of Alpena, Michigan. In 1890, Alger secured timber limits from the Ontario government near Blind River, in Cobden township, on the north shore of Lake Huron. Logs for his mill were rafted from his timber limits across Lake Huron to Black River.

The Dingley tariff bill, having the support of the newly elected Republican William McKinley (president, 1897-1901), who had gained his party's nomination because of his support of higher tariffs, became law in the spring of 1897. In reinstating the two-dollar per thousand feet rate on lumber, the Dingley tariff broke what many Canadian lumbermen had taken to be a sort of gentleman's agreement on trade. But the legislation went even further: it aimed to prevent the Canadians from reimposing an export tax on saw logs through inclusion of a countervail provision: the American tariff could be increased by an amount equal to any levy that Canada might impose. The effect of the tariff was immediate. The migration of sawmills was halted, and the Michigan lumber industry had a new lease on life; mills there, like Alger's, protected by a tariff wall and a threat of reprisal, were able to saw logs brought down duty-free from Canada.

The Dingley tariff immediately led to closure of many marginal Georgian Bay operations and severely reduced production by most of the large mills. The Georgian Bay Lumber Co., for example, reduced the cut in the Waubaushene mill from 30 million board feet to 26 million in 1897 and to 24 million in 1898. Adding insult to injury, Michigan lumber mills were sawing Canadian wood for the American market while Canadian-produced lumber was penalized in the same market. The Canadian government was powerless to act because of the countervail provision.

Canadian lumbermen, displaying the kind of ingenuity of which they were always capable, and acting together in the face of the common danger, found a solution themselves. If the Dominion government was blocked on any tariff action, why could the province, which owned the resource, not act? Could it not, through its timber regulations stipulate that sawlogs cut on crown lands could not be exported? Persuaded by the lumbermen to enact such a regulation, this is exactly what the Crown Lands Department did. On 17 December 1897, crown timber regulations were amended, with a provision that all pine logs cut on crown-licensed timber limits or berths had to be manufactured into lumber or timber in Canada. A bill, approving the regulation change, was rushed through the legislature and was assented to on 17 January 1898. The act[9] was immediately challenged by American limit holders in Ontario, on the grounds that it interfered with exports and was therefore ultra vires provincial authorities, but the courts upheld Ontario's action.

Rarely has there been a clearer example of being hoist with one's own petard. Many Michigan lumbermen[10] closed their mills, built new ones in Canada, and paid duty under the Dingley tariff which they themselves had created.[11]

Instead of injuring the lumber trade, the Dingley tariff — and the subsequent action of the Ontario government — gave it a period of renewed prosperity. The year after the Ontario law was passed, lumber prices doubled. By the turn of the century, a period of great expansion had begun in Canada and a sizeable home market had developed. Settlers began filling the western plains, and the vast region between the Red River and the Rocky Mountains, which in 1899 was only thinly populated, by 1905 boasted three new provinces. Montreal and Toronto became quite large cities, and many smaller communities grew to centres of importance.

The vigorous economic expansion, as might be expected, benefited the Canadian lumber trade. The years between 1900 and 1914, except for the short depression of 1907-8, were a period of great prosperity for the trade. Prices increased steadily,[12] and demand for lumber in both domestic and US markets remained strong.[13]

With uncanny ability to forecast and the fortitude to take risks, Sheppard ensured that Georgian Bay Lumber had the resources to ride the wave of prosperity. Many of the timber limits held by the firm since the 1870s were becoming rapidly depleted; the pine in the Severn River limits was already exhausted. When Georgian Bay Consolidated Lumber had been liquidated in 1893, 760 square miles of timber limits had been sold to Merrill and Ring. These included all the timber berths previously owned by the Maganettewan Lumber Co. in Parry Sound District and 162 square miles of pine in Burwash, Maclennan, Scadding, Servos, and Tilton townships[14] in Sudbury District.

If the company were to take advantage of the expected boom in the lumber trade, clearly these reserves of timber would have to be replaced; consequently, Sheppard decided to invest a large amount of capital to obtain more pine. He bought in two areas in the north where the bulk of virgin pine still existed — the Blind River area of Algoma District and in Sudbury District, around Wanapatei Lake, twenty miles northeast of the developing mining town of Sudbury.

By 1900 timber licences had been issued for most of the accessible pine in Algoma, and so Sheppard was obliged to buy from private individuals. In 1900, he bought two thirty-six-square-mile timber berths from Cook Brothers of Midland — township no. 161, located northeast of the village of Blind River and berth no. 194, in Otter township, northwest of Blind River and adjacent to the Mississagi River. In 1903, he bought township no. 162, also from Cook Brothers.

H.H. Cook who, it will be recalled, bought extensive timber limits in Muskoka, in partnership with Anson Dodge, at the timber auction of 1871, proved to be as foresighted as Sheppard. In the sale of 1872, Cook bought, purely for speculative purposes, the timber limits on the north shore of Lake Huron that were sold to Sheppard thirty years later. Berths no. 161 and no. 194 had cost Cook only $4,480 each, while no. 162 had cost him $6,660; he sold them to Sheppard for substantially more. Because it was a private sale, no record of the price exists: only the licence transfer was recorded in government documents. But, given the average price per square mile of timber in 1903 ($7,000), Sheppard probably paid Cook about $250,000 for each of the three berths. Even with the

$72 ground rent per berth, paid annually to hold the licences, Cook made a handsome profit — well over $700,000.

Earlier, in the timber sale of 1897, Sheppard had bought two small timber limits — 15¼ square miles in area — in Kelly township, on the east side of Wanapitei Lake, for $32,475; in 1899, he bought nine more square miles in Bowell township, west of the lake, for $11,700. In the crown timber sale of 1903, he bought five more berths: in Alymer, Mackelcan, and McCarthy townships, north and east of Wanapitei Lake. The latter berths totalled forty-two square miles and cost $289,500. Georgian Bay Lumber then owned most of the valuable pine land surounding Wanapitei Lake.

When Sheppard bought these berths, he was not yet president of Georgian Bay Lumber; he had to obtain approval for the purchases from Josephine Dodge, who still held the majority of the shares and was president in all but name. As pointed out in chapter 12, Josephine approved the purchases, even though they meant she received no dividends for a year or two, preventing her from windingup Arthur's estate on schedule.

And so, in a space of about four years, Sheppard bought, with Josephine's and the other directors' approval, over $1 million worth of timber limits. Such a large investment took pluck and considerable confidence in his own judgment, but because of his daring act, the company was assured of an adequate supply of pine. The logs had to be driven many miles from the berths to Georgian Bay and boomed the full length of the Bay, but the Waubaushene mill always had a continuous supply of logs and was able to run at full capacity every year, generating huge profits for the shareholders.

Sheppard's timber investments were not limited to those he made on behalf of Georgian Bay Lumber. Just as William Earl Dodge had established lumbering operations, independent of Phelps Dodge, to provide business opportunities for his many sons, Sheppard too bought timber limits and established independent companies, for the benefit of his five sons and his son-in-law, Fred Gray. In 1899, he bought 12¼ square miles of timber in Hunter township in Algonquin Park.[15] He built a stcam sawmill at Brulé Lake station on the Canada Atlantic Railway (originally the Ottawa, Arnprior and Parry Sound Railway) which had been completed through to Depot Harbour (Parry Sound) in 1897.

He then incorporated the Sheppard Lumber Co., capitalized at $85,000, with 850 shares of $100 each. His son Charlie and his

Sheppard Lumber Company saw mill, Brulé Lake, Algonquin Park c1900 — courtesy Gerald Gray

son-in-law owned, between them, about two-thirds of the stock. Sheppard kept 182 shares, and the rest were divided among his daughter, Annie (Gray's wife), his daughter-in-law, Ellen (Charlie's wife), and two other sons, Leigh and Stanley. Sheppard was president, and Charlie was secretary treasurer. Production started in 1899, with Charlie in charge of the mill and overall management of the company; Gray superintended the bush operations. Young Leigh and Stan Sheppard, and probably some of their Gill cousins also worked in the mill.

Charlie, oldest of the Sheppard children, was born in Gravenhurst in 1876. After finishing business college in Belleville he was taken into Georgian Bay Lumber to learn the lumber trade. Following in his father's footsteps, he worked in the store at Port Severn, scaled in the bush, and spent some time in the mill and office at Waubaushene. In 1898 he married Ellen, sister of Charlie Stocking, who, with her widowed mother, Caroline Peckham Stocking, had moved to Waubaushene from Newmarket, in the early 1890s, to keep house for Charlie. The wedding took place in the Dodge summer home. Charlie and Ellen raised four children.

Fred Gray was born in Coldwater in 1878, son of Scottish pioneer settler John Gray. Fred's mother operated a drygoods store in the village. In 1898, he married his cousin, Annie Sheppard, only daughter of W.J. and Ellen Sheppard. Fred and Annie had six children — four boys and two girls. It is not known whether Fred worked for Georgian Bay Lumber before moving to Brulé Lake, but earlier he had somehow learned the art of timber cruising and scaling; he was involved in these activities until his premature death of cancer in 1920.

Fred and Charlie took their young wives to Brulé Lake with them; their oldest children spent their first few years in somewhat primitive accommodation in the tiny village that grew up around the mill and the station.

Some of the lumber sawed at Brulé Lake was shipped all the way to Depot Harbour on Georgian Bay for transshipment by freighter to Great Lakes ports, but most of it was transferred to the Ontario Northern Railway (now CNR) at Scotia Junction and shipped to Toronto. The company also operated a hardwood mill and stave factory at Maple Lake Station, on the Canada Atlantic Railway, about eighteen miles east of Parry Sound.

Charlie and Fred ran such an efficient operation at Brulé Lake that the small timber limit was cleared of pine in four years. The timber licence was allowed to lapse, and the mill was subsequently sold. Charlie Sheppard moved back to Waubaushene and eventually became superintendent of Georgian Bay Lumber. Annie Gray established a permanant home in Coldwater, near her mother, while Fred went off to work in the bush, like his father-in-law before him.

In 1907, the Sheppard Lumber Co. bought two new timber berths in Algoma District from the Western Bank of Canada — all of Proctor township and the adjacent, unnamed township, no. 143. These berths were located a few miles east of the present-day uranium-mining town of Elliot Lake. Fred Gray and Leigh Sheppard were in charge of logging operations. Logs were floated down the Serpent River to Georgian Bay and boomed to Waubaushene, where they were sawn under contract to the Georgian Bay Lumber Co. A flag station and railway siding called Sheppard (still marked on road maps) was established at the point where the Canadian Pacific Railway crosses the Serpent River, at the spot where the logs were driven into booms for the long trip to Waubaushene. Men and supplies destined for the lumber camps were unloaded at Sheppard.

Fred spent most of his time in the bush camps. In 1911 he joined the newly chartered Masonic Lodge in nearby Spragge, so presumably he spent some time in that village as well. During the summer months, when school was out, Annie and the children joined Fred at Sheppard, where they passed the summer holidays in a tar-paper-covered shack belonging to the company.

These timber limits, which must have been partially cut before the Sheppards bought them, were exhausted by 1912. The licences were then sold to Traders' Bank. Fred Gray moved back to Coldwater, and Leigh was transferred to Sturgeon Falls, where he supervised Georgian Bay Lumber's logging operations on the Sturgeon River. In 1909 the Sheppard Lumber Co. purchased 11¼ square miles of timber in Shawanaga township, near the Shawanaga Indian reserve on Georgian Bay, and a few miles north of Parry Sound. Logs from this berth were also sawn in Waubaushene.

Also in 1909, when the lumber market was buoyant and prices had reached all-time highs, the Sheppard family established three more lumber companies: the Dunn Lumber Co., in partnership with Jack Dunn and W.H.F. Russell; the Sheppard and Dunn Lumber Co.; and the Muskoka Lakes Lumber Co. As far as can be ascertained, the first two firms did not own timber limits in their own right; they seem to have been primarily wholesalers of lumber,

Sheppard Lumber Company camp near Spragge c1910
– courtesy Gerald Gray

Cook House, Sheppard Lumber Company camp c1910
– courtesy Gerald Gray

acting as agents for Georgian Bay Lumber and other companies. Muskoka Lakes Lumber, which was owned jointly by W.J. and his son Charlie, had timber limits in Cardwell township, in Muskoka, at the north end of Lake Rosseau. The company produced lumber in a mill at Rosseau. Stan Sheppard managed the operation for his father and brother between 1909 and 1912; then he moved to Toronto to enter the stock-broking firm of T.S Richardson.

Sheppard's principal interest, apart from business and politics, was his family. He ensured that his sons and his daughter received good educations. At one time or another, he employed some of his sons on the Georgian Bay Lumber staff, later providing the capital to launch each in his own business career, be it farming, stock-broking, or lumbering. He adored his eighteen grandchildren; like most grandfathers he was more tolerant of their foibles than of the errors of his own offspring. He provided a home and employment to his niece Maud Skene, as companion and housekeeper to Ellen and after her death, as housekeeper to himself.

The Sheppard lumber companies ceased operating when the Waubaushene mill went out of business; all charters were surrendered in 1924. Because the Sheppards owned and operated the companies, the large profits generated during the hey-day of the trade remained in the family, giving the boys the capital necessary to pursue other businesses. Leigh Sheppard and his brother Herbert, in partnership with their cousin Ernest Gill, established a retail lumber and coal business — the Sheppard and Gill Lumber Co.[16] — in Toronto, in 1922. Annie had enough money from the Sheppard Lumber Co. to move to Toronto and to give all her children university educations. Charlie left the employ of Georgian Bay Lumber for a time but rejoined it again as secretary treasurer, when Charlie Stocking, at the age of fifty, resigned to join the army during the First World War.

W.J. Sheppard entered into other lumbering partnerships in which none of his immediate family was involved. He and his brother Tom bought 100 shares each in the Metagama Lumber Co., which was chartered in 1899. Little is known about this firm's activities, but it apparently owned timber rights near the CPR flag station of Metagama, on the Spanish River, some one hundred miles northwest of Sudbury. The company folded in 1904 when the president, George McCormick, died.

In 1893, Sheppard formed a partnership with William Irvin of Peterborough. They bought thirty-one square miles of timber limits in the Rainy River area of northwestern Ontario, but plans to lumber there collapsed and the limits were sold the following year.

One lasting and highly profitable partnership was formed with Peter W. Wallace of Midland in 1909. Wallace had been sent by Sheppard to the Blind River area in 1899, presumably to search out viable timber limits for Georgian Bay Lumber. Subsequently, the company bought the three timber berths from Cook Brothers of Midland discussed above, and Wallace supervised logging operations for the next few years. To facilitate logging on the Blind River system, lumbermen established three transportation and slide companies, in which Georgian Bay Lumber held stock: the Blind River Transportation Co., the Blind River Boom and Slide Co., and the Blind River Driving Co. Blind River Transportation operated a series of steam tugs, alligators, barges, and narrow-gauge railways for carrying freight, men, and horses up the Blind River system to the several timber berths located on it. Lumbermen paid fees for the service. Blind River Driving and Blind River Boom, one operating upriver from the other, had authority to build water control dams and to levy tolls on logs sluiced down the river to the North Channel water. Wallace sat on the boards of directors of all three companies, as Georgian Bay Lumber representative.

So efficiently ruthless was the harvesting of timber in the early 1900s that a thirty-six-square-mile timber berth was normally cleared of pine in about ten years. (Georgian Bay Lumber licences covered only pine. It never did get into the business of sawing less exotic woods, such as hemlock and spruce.) By 1909, it was clear to Wallace (and Sheppard) that the company's timber on the Blind River would last only a few more years.

Consequently, Wallace left Georgian Bay Lumber and formed a lumbering partnership with Sheppard and J. McCormack. They bought timber berths in Scarfe and Patton townships, slightly north and west of the village of Blind River. It is not known whether they contracted to have their pine sawn in a local mill or whether it was boomed to Waubaushene, along with Sheppard Lumber and Georgian Bay Lumber logs, but the partners did not own their own sawmill. They logged until 1924, when the timber licences were sold to the Royal Bank. Meanwhile, in 1912, Georgian Bay Lumber's pine in the Blind River area having been exhausted, the limits were sold to John H. MacDonald of Bay City, Michigan.

In addition to the timber limits that Sheppard bought for his and his sons' lumber companies, he had to be constantly on the look-out for new supplies of timber for Georgian Bay Lumber. For his own small companies to remain productive, the main mill at Waubaushene had to be kept viable. Moreover, as president of Georgian Bay Lumber, he was accountable to the other shareholders and directors, who owned 77.5 per cent of the stock; they expected him to keep their company profitable, too.

Clearly, he could not buy all the available timber limits for his own use, even if he could have afforded to. Timber licences were becoming extremely expensive, increasing in value at every timber auction, as the price of lumber went up. Furthermore, Michigan lumbermen, desperate for logs, tended to push up the bids for limits at provincial timber auctions. Also, the ground rent was increased from $2 to $5 per square mile to discourage companies from buying timber limits for speculative purposes. Licence-transfer fees were also increased from $2 to $5 per square mile. All of these changes increased provincial revenues but added substantially to the cost of producing lumber.

A survey of the records of crown timber licences, issued and transferred in Ontario, shows that Sheppard bought limits with a view to maintaining about ten years' supply of timber on hand. Thus, as timber berths in one area were cleared, loggers could move into new districts, and a continuous supply of logs could be delivered to Waubaushene, year after year. One or two years before loggers moved into an area, a small gang was dispatched to build roads, improve rivers and streams, and construct logging camps.

The ceaseless search for new timber lands required Sheppard to maintain a continuous watch over the lumbering scene in Ontario. He kept an eye out for companies going out of business and willing to sell their timber rights at a good price. He maintained close vigilance over the activities of the Crown Lands Department — later the Department of Lands, Forests and Mines — and was generally aware in advance when new townships were slated for public auction. Nothing was left to chance. He kept timber cruisers constantly surveying prospective licence areas, so that when timber berths were offered for sale he knew exactly what was in them and how much they were worth. The most reliable cruisers were Charlie Sheppard, Fred Gray, and Fred's cousin, George Gray.

The last provincial timber Sheppard bought for Georgian Bay Lumber was located in McWilliams and Thistle townships, twenty miles north of Sturgeon Falls and a few miles southwest of the Marten River Provincial Park, and in Blyth township, ten miles northwest of North Bay. Logs from these berths had to travel 150 hard and expensive miles from the skidways to the Waubaushene mill. Logs from McWilliams and Thistle were floated down a network of lakes and streams into the Sturgeon River and driven to Sturgeon Falls; from there they were towed across Lake Nipissing to the outlet of the French River, then driven down the river to Georgian Bay by the French River Boom Co. From the mouth of the French River, they were towed in booms to the mill by the company's steam tug *Waubaushene*. Logs from Blyth township were driven down the Little Sturgeon River to Lake Nipissing.

The last and most expensive timber purchased by Sheppard was from the Indians of the Dokis Reserve. We discuss that purchase in the next chapter.

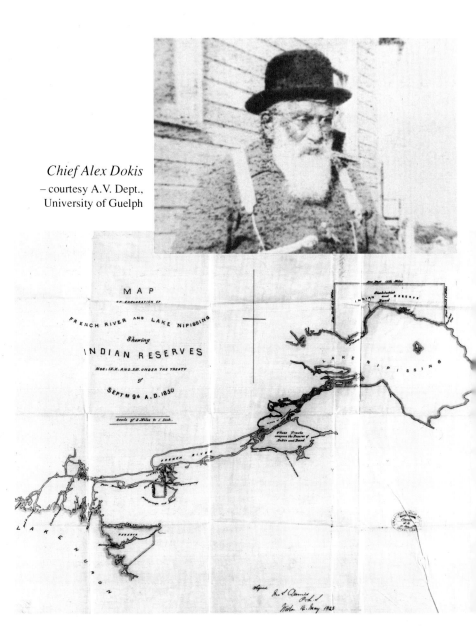

Chief Alex Dokis
– courtesy A.V. Dept.,
University of Guelph

– National Archives of Canada

244

Chapter Fourteen

The Eagle and His Band

The current tendency of native people and naturalists to resist logging operations in the few remaining stands of virgin timber in Canada is not a new phenomenon. Over a century ago. Chief Michel Dokis, leader of a small band of Indians, refused to allow lumbermen to cut the valuable timber on his tiny reserve on the French River. Whereas other Indian bands in the north were willing to surrender their timber for an amount equal to the sale price set by the Ontario government, Dokis and his band, for over thirty years, prevented the Dominion government from selling their timber to anyone. Dokis's continued obstinacy in the matter made him the nemesis of lumbermen, Indian Affairs officials, and politicians, all of whom tried various forms of argument, bribery, threat, and deception to pry the timber away from the chief's control.

By 1900 the grim destruction of the last of the stands of virgin white pine that had once covered eastern Ontario was well under way; of the pine there remained only scattered pockets in Muskoka and Parry Sound districts and a narrow strip extending from Mattawa on the east to Sault Ste Marie on the west. The deadly assault on what was left of the northern flank of the boreal forest was highly organized and efficient. The destruction of the forest was immense, and the consumption of logs by sawmills on the north shores of Georgian Bay and Lake Huron enormous. Every year during the first decade of the century, 700 to 800 million board feet of pine lumber was produced by the mills on Georgian Bay alone. Of this total, the Georgian Bay Lumber Co. accounted for an annual average cut of 30 million feet.

By 1920 the slaughter of the forests was nearly over, and only small pockets remained to be mopped up. Just over fifty years had elapsed since Anson Dodge bought timber limits on Georgian Bay with the expectation of "coming through and being wealthy," from this valuable provincial resource. There had never been any thought of conservation or renewal.

While the cutting lasted, a few lumber barons became rich. Even Anson Dodge would have been amazed at the wealth generated by the forests that he once owned. At a conservative estimate, the annual profits from lumber produced on Georgian Bay amounted to $15 million, assuming an average profit of $20 per thousand feet. Georgian Bay Lumber profits averaged between $700,000 and $800,000 per year, with some years even higher. Though the lumbermen had the strongest motives for exploiting the forests, they were not the only ones to benefit. Indeed, just about everyone in Ontario benefited to some degree from the lumber trade; hence few questioned the devastating assault on the northern forests.

Consumers valued the product. White pine was the pre-eminent soft wood and highly sought after. It was soft and easily worked, yet strong and light, and it could be used for nearly every wood-working purpose: building timber, ordinary lumber, flooring, doors and windows, furniture, decking, masts for ships, and engineers' patterns. Builders and contractors could never get enough of this splendid material.

Thousands of lumber-jacks and mill workers depended on the Ontario lumber industry for their livelihoods. Though the material wealth from lumber that grew so spectacularly during the early years of the century was not distributed equally — wages never kept up with dividends — the bouyant trade did provide steady employment. Unions were non-existent, but loggers occasionally balked at the low pay and refused to enter the woods unless their wages were raised. (Sheppard's election promise in 1911 was a bit of a sham. He would have had to raise wages no matter who won the election; otherwise, he would not have been able to attract his share of the Quebec lumber-jacks who came up to Ontario every year and on whose labour the industry depended heavily.) But if wages were not high, they were at least comparable to those in other occupations. If a bushworker avoided losing his money in the nightly camp poker games, he could emerge from the bush in the spring with a respectable sum. Besides, many young men liked the free-wheeling, macho-style life that lumber-jacking offered.

Northern settlers also profited from the industry. Some found employment in the lumber camps and mills, but. more important, the sale of timber from their patented lands to lumber companies provided cash incomes — for some, the only source of money in the winter. During the Dingley tariff controversy, most settlers opposed any measure that would interfere with the sale of sawlogs to American exporters, who generally paid higher prices than Canadian operators.

The exploitation of the forests was supported whole-heartedly by municipal leaders in the north, for in most cases the industry was the sole basis of a community's economic development and frequently the primary reason for its existence. Pine bought progress, attracting residents and secondary industries and paying for town halls, churches, libraries, community halls, and hockey rinks.

The provincial government, and through it the whole province, benefited from the northern lumber trade. Provincial revenues from timber sales, timber dues, and ground rent increased steadily every year; by 1901 these revenues amounted to $1,419,847, the largest sum ever. The director of forestry was able to report to the legislature: "Receipts from Crown Lands [were] by far the most condiderable source of Provincial revenue, with the exception of the Dominion subsidy."[1] In 1901 the total revenue collected by the Crown Lands Department, including income from the sale of agricultural lands, mining licences, and miscellaneous sources, amounted to $1,634,724. Expenditures were only $252,977, leaving a surplus of $1,381,747 — most of it accruing from the lumber industry — for politicians to spend on their pet projects in the south. It would be many years before other sources of provincial income exceeded that from the forests.

One of the few men in Ontario to see no virtue in destroying the forests was Chief Michel Dokis. Chief Dokis was born somewhere near Lake Nipissing in 1818, the son of a French-Canadian fur trader, Michael d'Aigle, and his country wife, Louise Obtagashio, a full-blood Ojibwa Indian. D'Aigle left Louise after the birth of her first child, Michel, and shortly afterward she joined another French-Canadian, Michel Restoul (Wahshusk), by whom she had two more sons, Francis and Joseph Wahshusk (Restoul). She kept Michel, who later became Chief Dokis, with her, raising the three half-brothers together. Louise never returned to her first mate, but continued to live with Restoul until his death, and afterward with Michel.

Dokis's Indian name was Migisi, Ojibwa for "eagle," which was the English translation of his father's name, Aigle. The name Dokis was a sobriquet, a corruption of the English word for ducks, which was applied to Michel as a young boy because of his inability to pronounce the plural of duck plainly, calling them duckies. Although he and his sons went by the name Dokis, the chief usually signed letters with his full name, Michel d'Aigle Dokis.

Dokis was an enterprising young man. He and his half-brother, Francis Restoul, became traders, operating a post at Dokis Point, on the north shore of Lake Nipissing. For some years they also traded on Lake Temagami, in opposition to the Hudson's Bay Co. When that company temporarily closed its post at Temagami Island in 1865, Dokis moved in for a time, hiring Indians to bring his trade goods up by canoe from Lake Nipissing.

Although half-breeds, Dokis and his half-brothers were raised as Indians; but, whereas the Restouls learned English, Dokis became only minimally proficient in the language and had no formal schooling. After visiting Dokis in August 1898, a timber inspector with the Department of Indian Affairs described the chief as "straight, active and well preserved, of benign and refined aspect. He is a sensible man of large experience among Indians and White people."[1] The same official judged him to be "thoroughly upright and candid in his statements, and appreciates these qualities in others."[2] A superintendent of Indian Affairs who several times tried unsuccessfully to wrest the chief's timber away from him was less charitable. He observed that Dokis exercised an influence over his band that was "of a tyrannical nature" and that he wielded "absolute control over his relatives,"[3] who formed a considerable portion of the band.

In truth, Dokis was not only chief of his small band, but father and uncle of it. Originally, it had consisted exclusively of his sons and daughters and their offspring; later, it included his Restoul half-brothers and their families. It seems that he was a self-proclaimed chief; as an Indian Affairs official put it, "Chief Dokis . . . naturally assumed chieftainship of this small tribe, owing to his masterful character."[4]

Self-proclaimed or not, Dokis was recognized as a chief by the Hon. William B. Robinson, when the latter met with the Ojibwa at Sault Ste Marie in September 1850 to negotiate the surrender of a large block of Indian territory, extending from Penetanguishene to Batchawana Bay on Lake Superior. In return for the surrender, each band was granted a reservation, to "be held and occupied by the said Chiefs and the Tribes in common for their own use and benefit."[5] They were also guaranteed hunting and fishing privileges on the land they surrendered and were to receive an annuity of one pound (four dollars) a head.

Dokis, who signed the surrender treaty on behalf of his band, knew precisely what piece of land he wanted for his reserve, even though, in 1850, no one in the band lived on it. He selected an island in the French River known as Okikindawi — "home of the buckets" — and an adjacent peninsula, joined to the mainland by a narrow neck of land only a quarter mile wide. The reserve, consisting of about sixty-one square miles, was confirmed and registered in 1853 after a survey by PLS John S. Dennis, conducted under the watchful eye of Chief Dokis.

The chief had chosen one of the most valuable pieces of real estate in northern Ontario, although, as subsequent events showed, he had not chosen it for its commercial value. The small reserve contained one of the finest stands of red and white pine, hemlock, and assorted hardwoods in the north. The pine was plentiful, of exceptionally fine quality, and generally free from such defects as spunk and hollow butt; a fair proportion of it was large in diameter, long and clean, and suitable for square timber. The ground was level, and no point on the reserve was more than a mile and a half from either the French River or its tributary, the Memesagamesing River, which formed the southern boundary of the reserve. Naturally, such a valuable and accessible piece of timber land was soon coveted by lumbermen. And in fact both the Robinson-Huron Treaty and the subsequent Indian Act permitted the Dominion government to sell or lease timber or minerals on reserves on behalf of the Indians, but only if the Indians agreed.

The first official application to cut timber on the Dokis Reserve was made in 1881 by Richard Power of Parry Sound. In the next five years, the Department of Indian Affairs received no fewer than thirteen more applications from lumbermen, many of whom had first approached Chief Dokis. The first serious attempt by the government to obtain the surrender of the timber was made in 1886, when the superintendent general of Indian Affairs instructed Thomas Walton, regional Indian superintendent, to arrange through the chief for a surrender vote by band members. Walton met with the chief and explained to him the advantages of surrendering the timber: interest from the invested bonus money and subsequent timber dues would guarantee every member of the small band a permanent income of at least four dollars a year; and rapid settlement of the district placed the pine stands in great danger from fire. After listening patiently to Walton's arguments, the chief replied that he was averse to selling the timber. As for the possibility of losing it to fire, he told Walton, "If it [is] the will of Providence that it should burn then let the timber burn."[6] Walton assured the chief that he respected his personal opinion on the matter, but by the terms of the Indian Act he had to get the opinion and vote of the whole band. For this purpose, he requested the chief to hold a full meeting of the band on the next annuity pay day.

Annuity pay day took place at the end of September. After making the annuity treaty payments, Walton delivered a long, patronizing lecture, through an interpreter, to the band members. He drew their attention to the danger from fire, he emphasized the large amount of interest that would accrue to each member of the band if the timber were sold, and he lectured them about their duty to place their inheritance and that of their children in the safety of an interest fund. He reminded them that all the other bands in the district had surrendered their pine and were receiving substantial annual interest payments.

But it was all to no avail. The band members "half indignantly and unanimously refused to entertain the subject of surrendering their timber and seemed hurt at being so often approached on the matter."[7] "Unfortunately," an incredulous Walton reported to the superintendent general,[8] "my arguments were in vain, and I was forced to the conclusion that sentimental rather than monetary considerations guided their conduct."[9]

Sir John A. Macdonald, who was under considerable pressure from lumbermen to take action, toyed with the idea of introducing an amendment to the Indian Act "enabling the Governor in Council to deal with timber reserves when tribes [were] unreasonable."[10] But on further reflection, and remembering the outcry a year and a half earlier following the hanging of the Metis leader, Louis Riel, Macdonald prudently decided not to act.

An alternative possibility was suggested by the deputy superintendent of Indian Affairs, L. Vankoughnet, who noted that an 1885 amendment to the Indian Act granting the superintendent general authority to issue licences to cut trees on reserves did not include a provision for prior Indian approval. He wondered if this clause might be a loophole that would allow the government to grant licences to cut timber on the Dokis Reserve without the consent of the band.

The opinion of the department's lawyers, whom Macdonald consulted, was that the amendment only clarified the matter of issuing timber licences as one means of disposing of timber owned by Indians. Neither the Robinson-Huron Treaty nor the subsequent Indian Act had specifically used the term *licence* referring instead only to the Indians' rights to sell or lease timber on the reserves. Though the amendment permitted the superintendent general to issue licences without the approval of Indians, he could do so only after the Indians had previously agreed by a vote to surrender the timber. Clearly, the clause in the Indian Act could not be used to pry Chief Dokis's timber away from him without his and his band's consent.

When Walton was again asked for his opinion a year later, he concluded that the Indians would never release the timber voluntarily because of the chief's attitude. It seems that Dokis "had promised to the Hon. W. B. Robinson, in 1850, that so long as he, the Chief, lived the Timber of the Reserve should not be sold."[11] The chief considered this a sacred oath which he was determined not to break. Walton reported that while the chief and some of his immediate relatives were fairly well-off, some members of the band were poor; yet "the obstinacy or sentimentality of the Chief yearly deprives all such of $20 of Interest money."[12] Consequently, he recommended "that the Timber on this Reserve be sold even though the Band has repeatedly refused to surrender the same."[13]

There the matter stood for another few years. Meanwhile Dokis moved two of his sons onto the previously uninhabited reserve, presumably to serve as watchmen, and he personally made a tour of the reserve every year "for the purpose of seeing that no deprivations [were] being committed and that all [was] right."[14] By 1893, intense lobbying for licences to cut on the Dokis Reserve began. Many applications came from Michigan lumbermen who, during the brief period of Canadian-US free trade in wood products, were removing large quantities of sawlogs, free of export tax, to their mills in Michigan. The Department of Indian Affairs, still hoping to come to terms with Dokis and his band, decided to have a formal valuation survey made of the timber on the reserve. Vankoughnet counselled extreme caution: "There is just the risk," he warned, "that if the Chief gets wind of the timber being inspected, as he is pretty sure to do, it may make him all the more determined not to surrender."[15]

Two evaluation surveys were conducted in the spring of 1893: one by an independent timber cruiser, P.H. Colton; the other by Indian Affairs' own timber inspector, George L. Chitty. Both confirmed that the Dokis Reserve was, indeed, a store-house of valuable timber, and both Colton and Chitty estimated the amount of marketable white pine on the reserve at 45 million board feet. It was judged to be much superior in quality to the timber then being cut on Lake Huron, for which purchasers had paid very high prices. Judging from what crown timber limits near the reserve had sold for, and given Michigan lumbermen's interest in the timber, it was reckoned that a public auction of the Dokis timber would net at least $250,000.

As predicted by Vankoughnet, Dokis did take exception to the surveys. In July 1893 he wrote to the deputy superintendent general complaining about the "number of persons that comes and walks our reserve to see the wood and it puts of our beast [game] of our reserve." He reiterated that the band "cannot sell the wood, that is on our reserve, that we want the use of the wood and the children that are to come will want the use of the wood as well as us to build them houses or other local purposes." He begged the government "not to sell the wood to any lumbermen to cut." Finally, he asked that the surveys be stopped, because they "caused great damage to [them] that [they] can't see any beast on the land."[16]

The chief's perfectly legitimate complaints and reasonable requests were ignored; the timber on the Dokis Reserve was considered too valuable a resource to be left in the hands of sentimental Indians. And so Walton was ordered to try, one more time, to obtain a surrender. With an upset price for sale by public auction now set at $250,000, the government was prepared to sweeten the pot (with the Indians' own money) if they agreed to surrender the timber. Walton was authorized to offer each individual $100 upon the signing of a surrender agreement and a further 10 per cent a year from the interest on the balance. The surrender vote was to take place on annuity pay day, set for 28 July. (Ontario newspapers had already announced the forthcoming sale of the Dokis Reserve. The Ottawa *Evening Journal* described the reserve as "a rich plum" and, quoting lumbermen, claimed it was "one of the richest timber limits in either Ontario or Quebec, if not in the Dominion."[17])

In preparation for the vote, and to ensure the compliance of the Indians, Walton wrote to each adult male member of the band explaining the financial advantages of a sale. Assuming that the sale of the timber netted $250,000, the band could realize $8,000 a year in interest on the capital; this would increase gradually as ground rent and timber dues were paid by the purchasers. Since, according to the 1891 census, the band comprised sixty-one members, the $8,000 in interest "being divided equally among them would give to each a little over $131 every year,"[18] Walton pointed out. He included in the letter a detailed list showing how much each family would be entitled to receive annually. Chief Dokis, with seven dependants would receive $971; William Dokis, with a family of nine, $1,179; Frank Dokis, $786; Alex Dokis, $971; and so on. If the band preferred, one-tenth of the sale price, $25,000, or $409 for each member, could be distributed at the time of the sale. Such a choice would, of course, leave less in the capital fund, reducing annual interest payments to about $119 per person.

But Walton's clever attempt to influence the vote by holding out a promise of rich material rewards failed. Chief Dokis had done his homework too. Thus, when Walton raised the matter of the timber surrender at the annuity pay day festivities, he discovered that "the subject of the surrender of their Timber had evidently been previously discussed and their minds seemed made up as to how they were going to vote," and he "immediately saw that they were going to refuse to surrender."[19] Nevertheless, Walton pointed out the advantages of surrendering and asked for their objections to such a course. The answers, which Walton considered "almost too frivolous to mention,"[20] were: "After the pine was gone, they would have nothing left; their children might want the pine; although I promised them $100 per head yearly, on the sale of the pine by and bye I would come and pay them perhaps $10 and say that that was all [there was] coming to them; they would have to pay for Schools etc. etc."[21]

Biding his time, Walton adjourned the council until 2 p.m., meanwhile leaving the Indians a letter from the deputy superintendent general that they "might peruse it at their leisure to satisfy themselves that what [he] had told them had official sanction."[22] Later, when he reassembled them, they told him that "they had placed their votes in the hands of their Chief."[23] Upon inquiring of Dokis, Walton was informed that "he had voted against surrendering the Pine Timber."[24]

"The action of the Band in this matter exemplifies in a marked degree the incapacity of the Indians to manage their own affairs," Walton lamented. "Because of the stubborn waywardness of one old man, their Chief, they refuse to execute an act that would place all in most comfortable circumstances," he complained. Finally, "in view of the incapacity of the Dokis Band to exercise any judgment in the matter of the surrender of their timber," he arrogantly recommended that the department "seek or take the exceptional authority to dispose of their Timber without their consent or without previously having obtained from them the surrender of same."[25]

Dokis had good reason not to trust Walton or any other member of the government. Walton had always treated the chief condescendingly; for example, he had refused to send translations of his letters to Dokis, who was obliged to seek out someone who could read English and could translate his responses. Walton's successor, J.D. McLean, though pledging friendship, had refused to make treaty payments on the reserve, as custom dictated; he insisted in making the payments at the Indian village of Beaucage on the Nipissing Reserve, half-way between Sturgeon Falls and North Bay. Hunting and fishing privileges that were guaranteed to the Indians by the Robinson-Huron Treaty of 1850 had been denied them by the Ontario government after Confederation, and the Dominion government had done nothing to restore the ancient privileges.

In 1897 the department, suspecting that poor personal relations between Walton and Dokis were the cause of the chief's unwillingness to surrender the timber, sent an experienced Indian agent, Alexander McDonald from Pembroke, to negotiate a settlement. But McDonald, who presumably spoke Ojibwa and was known to enjoy the confidence of the Indian people, also failed.

In the winter of 1899 the inevitable happened. Loggers of the Hardy Lumber Co. of Alpina, Michigan, inadvertently trespassed on the Dokis Reserve and cut 1,129 pieces of the chief's timber, owing to an error by the company's foreman in determining the boundary between its timber limits in Hardy township and the adjacent reserve. When the company discovered the mistake, its Canadian representative immediately informed the superintendent of Indian affairs, not the chief. Without consulting the chief, the deputy superintendent general sent the department's timber inspector, George Chitty, to investigate. Chitty hired a scaler, licensed by the province, to measure the timber, and, though the scaler estimated the amount of pine cut at 172,336 board feet, the superintendent general charged the company only $1,044 for the logs and allowed it to haul them off the reserve and onto the company's log dump on the Memesagamesing River. No penalty was imposed.

When Chief Dokis learned about the trespass and the arrangement that the government had made, he was furious. He wrote to the Department of Indian Affairs expressing his displeasure; he rejected the $1,044 payment, asked for a more reasonable $2,040 in compensation, and demanded that the logs be "left on the ground to rot and by that check the desire of more people trying to steal timber."[26] The department, however, would not agree to the chief's demands. When he discovered that the logs had actually been hauled away, he again wrote to the department, threatening to retrieve the logs and to "have [the] parties that cut them arrested."[27] Indian Affairs' handling of the case was, typically, patronizing toward Chief Dokis and ingratiatingly sympathetic to the American lumber company. There is even some question whether the department had any legal right to act as it did.

Timber regulations made under the Indian Act with respect to trespass on Indian lands authorized the superintendent general to impose a "penalty equivalent to double, treble or quadruple the ordinary dues, "even if timber was cut "through error in good faith."[28] But this regulation applied only to timber cut illegally on Indian lands licensed to other parties. Clearly, because the timber on the Dokis Reserve had not been surrendered and consequently was not under licence, the superintendent general could not act under the regulation. Nor was section 26 of the Indian Act, which provided for a penalty, on conviction, of $20 "for each tree cut and carried away,"[29] much help either, for the logs had not been carried away, and the Hardy Lumber Co. had not been charged. Section 66 of the act allowed timber to be disposed of after one month if logs were seized. But in this case Chitty had not been ordered to seize the logs, although he did put a "broad arrow" on them — an indication that they belonged to the band.

If the superintendent general had no authority under the act or regulations to impose penalties, it would seem that he had no authority to sell the logs to the lumber company either. The logs belonged to the band and should have been turned over to the chief to dispose of as he wished, even if that meant leaving them on the ground to rot. It would appear, too, that since the logs were the property of the band, the chief should have been allowed to pursue the matter through the courts, as he proposed to do. Because of the government's generosity, Hardy had paid no more for the logs than if the trees had been growing on his own limits — an equivalent purchase price of $4 per thousand feet, and $2 per thousand feet in timber dues. The company was allowed to abscond with first-class pine, on which it earned a handsome profit.

Even though the department had acted arbitrarily and detrimentally to the Indians, and perhaps even illegally, the secretary of the department had the audacity to express to the chief "disapprobation of the general tone" of his letters, which were "couched in very unwarranted and disrespectful terms not only to the Minister but also to the Department."[30] Indian Affairs even placed the onus on the chief to protect his timber from future trespass; the secretary directed him to have a boundary line ten or twelve feet wide cut between Hardy township and the reserve. In disgust and resignation, the chief finally conceded to the superintendent general, "You are stronger than we are,"[31] and let the matter rest. But six or seven more years would pass before anyone in the department again dared confront the ageing chief on behalf of avaricious lumbermen for the surrender of his timber.

One lumberman who had had his eye on the Dokis timber for some time was Charles A. McCool of Mattawa, who in 1900 was elected to Parliament as the Liberal member for Nipissing. McCool used his influence with the Laurier government to have an acquaintance, George P. Cockburn, appointed Indian agent for the region, with headquarters in Sturgeon Falls. Cockburn's assignment was to obtain surrender of the Dokis timber, a fact confirmed by McCool himself in a debate in the House of Commons dealing with the proposed sale. McCool blatantly boasted to the House, "I was instrumental in getting the agent appointed as I had every confidence in him, and one of the reasons why I wanted him appointed was in order to get the surrender of this and another reserve."[32]

Cockburn, who seems to have understood Indian politics, knew how to exploit divisions in the band. The band was certainly divided on the matter of surrendering its timber. Most of the Dokises, supporting the chief, were against surrender, but the Restouls were in favour. Francis Restoul had not placed his vote in the "chief's hands" at the time of the 1893 decision.

Hoping to take advantage of the split, Cockburn spoke to each band member separately, and, by February 1905, concluded that "it might be possible to get a majority vote for the surrender of the pine timber."[33] When the deputy superintendent learned this, he asked

Cockburn for a list of the names of band members over twenty-one years of age who might agree to surrender, showing where they lived and the probable distance from the reserve of each non-resident. In obtaining the information, Cockburn was "to exercise caution in order to prevent premature publicity which would be sure to work serious opposition in the matter."[34]

When Cockburn began to talk seriously about the surrender, he discovered that only seven out of nineteen adult males would definitely vote yes; three of those lived at Abitibi, 180 miles from the reserve, and might not be able to attend a surrender meeting. Consequently, Cockburn was obliged to report that though he thought a surrender would be popular with the band, "they did not wish to vote against the wishes of Chief Dokis."[35] The chief, it seems, who was then eighty-seven years old, continued to exert "absolute control over his relatives."

On 25 April 1906, the old chief died. His last act as chief and father was to exact, from his death-bed, an oath from his son and heir that he "would hand down the timber to his children the same way as he himself had done, and that no disposition thereof by the Department should be consented to."[36] Dokis's oldest son, Michael, became acting chief on the death of his father, but before his position could be confirmed by election he too died. And so his brother, Alexander Dokis Sr, was elected chief on 11 July. Alex, however, was not capable of exercising the same kind of control over the band as his father, especially as regards the now very valuable timber.

Believing the change in leadership of the band, accompanied by jealousies and internal rivalries that so often occur at such times, provided an opportunity to get control of the Dokis timber, McCool exerted pressure on Frank Oliver, the superintendent general, to take action. Consequently, Cockburn was asked to investigate and to report "as to when he consider[ed] it would be a favourable time to bring again the question of surrender before the Indians."[37] He was instructed further to look into the value of the timber on the reserve "so as to reach an upset price in order that the Department [might] place it before the Indians when approaching them on a question of surrender."[38] Cockburn was also asked to determine if the Indians would be prepared to surrender their land as well as the timber.

The department was now willing to pay the Indians $100 each, immediately after surrender, and "a further sum sufficient to make up 10% of the amount realized from sale of the timber, the balance to be funded for the benefit of the band, and interest paid thereon."[39] If the Indians agreed to surrender their land as well as the timber, the department would make a distribution in excess of the 10 per cent of the sum realized from a sale. Because the "excess" would be paid out of the proceeds of the timber sale, the department naïvely hoped to get both the land and the timber for the money coming from the sale of the timber only. (This money, it must be remembered, would be the Indians' own money.)

In the event that the Indians were still averse to surrendering the timber and the land, the department intended to put the timber up for sale by public auction, "the same to be subject to surrender being given afterwards by the Indians." In this way it would be ascertained "what amount of money could be realized for the timber,"[40] the assumption being that when the Indians saw the large amount of money available, they would surrender the timber with alacrity. The proposition, however, was "not at this point to be brought before the Indians."[41]

Why the department thought it could justify selling timber over which it had no control, or how it planned to respond to purchasers if the Indians refused to surrender the timber afterward were not explained. The department's hope that the independent-minded Dokis Indians, most of whom now lived on the reserve, would voluntarily give up their land along with their timber reveals that these administrators were not only more artful than their predecessors but probably more ignorant of the Indian character as well. The department need not have feared that Cockburn would reveal the disingenuous strategy to the Indians. He was too shrewd for that; furthermore, he had a plan of his own design for securing a surrender vote.

Sensing that the time was now favourable for a vote on surrender, but knowing also that Chief Alex Dokis, bound by the oath to his father, was opposed to sale, Cockburn ignored the chief and negotiated directly and surreptitiously with the other Indians. When convinced that he had enough positive votes committed, he sought permission from the deputy superintendent general to call a meeting for 7 January 1908, to take a surrender vote. Notice of the meeting, which had not been previously discussed with the chief, was not delivered to him until 31 December. As the notice was in English, the chief had to travel all the way to Sturgeon Falls for a translation. By then, however, several days had elapsed, leaving him no time either to seek legal advice or to try to influence his fellow tribesmen.

Meanwhile, to ensure a congenial atmosphere at the surrender meeting, Cockburn arranged a feast the day before, on a point about a mile and a half from the reserve, and, contrary to the law, a large quantity of liquor was served. Cockburn's strategy worked. The next day he was able to telegraph the deputy superintendent general: "Your instruction of twenty-sixth December carried by good majority."[42] Eight Restouls and three Dokises voted for surrender; Chief Alex Dokis and five other Dokises voted against.

The department lost no time in securing an order-in-council approving the surrender and sale and arranging for a public auction. Chief Dokis tried to negate the vote, and, though he was not successful, he did manage to hold up the process by sending a sworn affidavit to the superintendent general registering his objection to the surrender on grounds that the Restouls, as half-breeds, had no right to vote and certainly no right to share in the proceeds of the sale. He also accused Cockburn of supplying liquor freely, "for the purpose of influencing the vote,"[43] and of engaging in other irregularities to procure the vote of the chief and other members of the band. The chief even went so far as to hire a lawyer in Orillia to submit a formal protest and to ask for a commission of investigation.

The chief's challenge of the right of the Restouls to vote was probably prompted by a family feud and his anger at their betrayal of the old chief's wishes, for the accusation had no basis in fact. The bloodlines of the Restouls and Dokises were similar; both families were half-breeds. William B. Robinson had acknowledged the rights of half-breeds to share in the treaty of 1850; otherwise he would not have permitted Chief Dokis to sign the treaty. Moreover, in 1874 the old chief had requested that his half-brothers, Joseph and Francis Restoul, be admitted to full membership in the band.

Accordingly, since 1875 the Restouls had been paid treaty money and had taken their full share and responsibility in elections and the ordinary business of the reserve. Their right to participate in the surrender of the timber and to share in the proceeds from the sale could, therefore, not be questioned. As for the liquor charge and the possibility that other irregularities had taken place, there was "no corroborative evidence of such on file";[44] the department had no intention of investigating these charges and running the risk of negating a surrender vote that had taken so many years to achieve.

It was later discovered that one of the principal reasons for the chief's strenuous objection to the surrender was that his daughter-in-law was expecting a baby, and he wanted to delay the sale until after the birth, thus ensuring that the child would be included on the pay list when the proceeds were distributed. The baby having been born, the chief withdrew his objection to the sale but continued to dispute the right of the Restouls to share in the proceeds.

The timber auction took place in the Russell House Hotel in Ottawa, on 27 June 1908. The reserve had been divided into eight small berths, ranging in size from 5.29 to 9.34 square miles. Each berth was offered for sale separately, at a bonus, 10 per cent of which had to be paid in cash on the day of the sale, with notes given for the remainder, payable in three, six, and nine months, with interest of 6 per cent. Purchasers were required also to pay timber dues of $2 per thousand board feet for sawlogs and $50 per thousand cubic feet for square timber, an annual ground rent of $24 per berth, and a $4 licence fee. The licences could be renewed yearly but were good for only ten years, after which all the timber would revert to the Dokis Indians. Only pine trees more than nine inches in diameter at the stump were covered by the licences.

A timber cruiser employed by the Department of Indian Affairs had determined an upset price for each berth, ranging from $50,000 for the smallest berth to $136,000 for the largest, and adding up to the incredibly large sum of $820,000. The auctioneer was given the upset prices just before the sale began; the bidders, of course, were denied them. At least they were not supposed to know what the minimum bids were. Some berths sold at, or slightly below, the upset price, but others sold for considerably more.

The total proceeds from the sale of the Dokis timber, lasting only a few minutes, amounted to $871,500. The bonus money, subsequent timber dues, and ground rent netted the band the amazing sum of $1.1 million. Thanks to the "stubborn waywardness of one old man," who had persistently refused to surrender his band's timber, Chief Dokis's descendants received not $4 per head per year, as originally offered, not $131 per year, as later promised, but a very substantial $50 per person per month. Thus, the Dokis Indians, then numbering eighty-one souls, became per capita the richest Indians in Canada. Not surprising, Charles McCool's bid for berth no.7 — the second smallest in area but the richest in pine — was accepted. The fact that he paid $131,000, $1,000 less than the upset price, suggests that he had advance information about the upset prices. (The berth was 6.8 square miles in area, and the upset price was $132,000. McCool paid $131,000.)

Another lumberman who was determined to get some of the excellent Dokis timber was W.J. Sheppard. As a Conservative, he had no political connections in the Liberal government in Ottawa to provide him with inside information, as McCool apparently had. Sheppard and representatives who bid on two other berths on his behalf simply bid until they got the berths Sheppard wanted. He bought three of the eight berths — nos. 2, 4, and 6. They totalled twenty-two square miles in area; the combined upset price was $301,000. Sheppard paid $334,000 for them, or slightly more than $15,000 per square mile.

Sheppard found lumbering on an Indian reserve frustrating, as compared to lumbering on Ontario crown land. First, he had to deal with inexperienced administrators in the Department of Indian Affairs, who knew little about the traditions and travails of the logging industry. Second, he had to contend with the Dokis Indians, who guarded their independence and their property with inherited sedulity.

Sheppard believed that the payment of annual ground rent protected the company's cutting rights on the reserve, as it would have on timber limits licensed by the Ontario government. He therefore decided, adhering to the company's practice of sequential clearing of limits, not to begin cutting on the Dokis Reserve for three or four years. He planned to finish clearing the company's limits in Sudbury and Nipissing districts, some of which had recently suffered fire damage and windfall, before moving his operations onto the Dokis Reserve.

Sheppard reckoned that the six or seven years that would then remain on the ten-year limit of the licences would allow ample time to cut the estimated 60 million board feet of pine on his three berths. It was with considerable shock and some displeasure, therefore, that Sheppard received a letter from the Department of Indian Affairs in June 1910 advising that the timber licences would not be renewed, because of failure to commence logging. The bonus sum of $334,000, paid for the timber, was to be forfeited.

It seems that the Dokis Indians, having refused for over twenty years to relinquish their timber, now liked being the capitalists the timber sale had made them. Because the Georgian Bay Lumber Co. was not logging their limits, annual timber dues were not being paid into the capital fund, denying the band interest on upward of $30,000 annually. The band pressed the department to take the timber away from Georgian Bay Lumber and sell it to some other company that would get on with the cutting.

"You have no right to withhold renewal of our licences," Sheppard started his letter of complaint. "The terms under which your Department sold these berths . . . gave us the right to cut this timber at any time within 10 years from the date of sale. We have carried out to the letter our part of the agreement, but your Department has not done this, but keep applying conditions from time to time that were not mentioned when we purchased, which we knew nothing about."[45] The department accepted Sheppard's arguments, and the licences were renewed.

Georgian Bay Lumber began logging on the reserve in the fall of 1911. By the end of the logging season the company had cut 9,666,971 board feet of sawlogs and manufactured 95,969 cubic feet of square timber, for which the band's capital fund received $20,780 in timber dues. Sawlogs were driven down the French River and towed to Waubaushene; square timber was rafted to Callander, loaded on the Grand Trunk Railway (now the CNR), shipped to Kingston for rerafting to Quebec, and eventually freighted to Britain.

Sheppard discovered extra costs imposed by the Indians that he had not known about when he bought the timber. On crown timber limits, logging companies used whatever logs were needed to build shanties and helped themselves to firewood for heating and cooking; no one worried about the value of this wood. Indeed, there was an unwritten understanding that this material was included in the bonus price of the limits. The Indians were not so understanding. Having surrendered only their pine, they insisted that Sheppard pay them for the hemlock used to build shanties and stables and twenty cents per cord for firewood.

In the summer of 1914 there was a small forest fire on the reserve. Although it was extinguished before any serious loss occurred, the fire underscored the extreme vulnerability of the company's large investment, as it had assumed fire risk when the timber was purchased. Sheppard recommended that the Indians be employed as fire rangers and paid out of the two per cent of the sale price that had been set aside for administration. The department and the Indians agreed, but Georgian Bay Lumber (and presumably the other companies) had to pay half the cost of the wages.

Another large forest fire burned through the company's limits in Aylmer, Mackelcan, and McWilliams townships in the summer of 1915. Since only the bark of a tree normally suffers immediate damage in a fire, much wood can be salvaged if the burned trees are removed immediately. Consequently, in the winter of 1915-16 Sheppard employed all his men and logging equipment to save the timber in these burned limits. No logging was done on the reserve, resulting in an interruption of Sheppard's cutting schedule.

By 1918 all the marketable timber had been cleared from the company's timber limits on crown land, so the licences were allowed to lapse. Only the Dokis timber remained in the company's possession. The Dokis limits were logged, as vigorously as wartime conditions would permit, in the winters of 1916-17 and 1917-18; but, by 30 April 1918, when the licences were due to expire and the uncut timber would revert to the Indians, there were still about 4 million board feet of timber on two of the company's berths. Sheppard asked for an extension of the licence for one year to enable him to cut this timber.

There were valid reasons beyond the company's control for its failure to clear the timber in the required time. The forest fire on the other limits was one of them. Also, in 1917 the Union government of Robert Borden passed the Military Service Act. Conscription brought loss of about 65 per cent of the company's normal complement of lumber-jacks, and reduction of its cutting capacity by about 400 per cent. Some unscrupulous lumber companies allowed draft dodgers, many of whom were French-Canadian loggers, to escape military conscription by working in the bush camps. But Sheppard, whose own son and his company's secretary treasurer, Charlie Stocking, were overseas, refused to "shield anyone from performing his duty to his country."[46] Consequently, he did not have enough men to harvest the timber on time.

The Indians opposed the extension and instructed the department to put the remaining timber up for sale again. This was not practical. There was not enough timber on the berths to justify the expense to another company of moving an outfit in to harvest it. The Indians themselves, as the department's timber inspector pointed out, "had no logging equipment of their own, nor had they the inclination or ability to log to any advantage."[47] Further, because the timber was surrounded by old cuttings and slash, there would be no chance to save it in the event of fire. If Georgian Bay Lumber's licences were not extended, there was a good chance that the timber and the income derived therefrom would be lost. And so the Indians agreed to an extension until 30 April 1919.

In the spring of 1919 Sheppard sought yet another extension. Timber inspector H.J. Bury, who had checked the berth and recommended the extension, advised the deputy minister that "the Company made every effort to remove the timber in the time specified, but owing to the outbreak of the epidemic of influenza ... were compelled to close down their camp."[48] About 1 million board feet of sawlogs remained.

Again the Indians opposed an extension. This time they pointed out something to the department that every Indian child knew: pine trees, like little girls, get bigger every day. Consequently, there was now timber on the reserve to which the company, under the ten-year term of the licence, was not entitled, because the original bonus had not covered it. After much haggling, a trait for which the Dokis Indians were by then well known and at which they were exceedingly adept, they agreed to extension only on condition that the company pay a bonus for the incremental growth of the timber.

From calculations based on the volume of timber and incremental growth, the department's timber inspector estimated that the remaining berth contained 100,000 board feet of timber that the company did not own. Because the incremental growth could not be stripped from the trees, the company was obliged to pay an additional $500 in cash or $5 per thousand board feet to make up the full stumpage value of the timber. The company also had to pay $2 per thousand board feet in timber dues for the additional wood. The timber was all cleared by the spring of 1920, when the licence finally expired.

The Dokis Indians prospered in the years ahead. For several years after the timber sale in 1908, each band member received a monthly payment of $50 from the timber fund. Indeed, compared to most Canadians, they were extremely well off, especially during the Depression. One of the Restouls, for example, who had twenty-one children, received an annual income of $13,800 throughout the 1930s.

In 1937, the band began receiving applications from local lumber merchants for the purchase of timber berths on the reserve, much of the pine left over from the earlier harvesting having now grown to marketable size. Again the band held a vote. Seventeen voted in favour of a sale; eight (all Dokises) voted against. Consequently, the band sold by tender 27 million board feet of timber — cedar, hemlock, maple, pine, and spruce. Three companies were issued licences, and three portable sawmills were erected on the reserve, at which some of the Indians found employment. Timber dues, averaging about $5,000 a year, were paid into the band's capital fund, which in 1940 still contained about $1 million.

Some capital was used in 1914 to build a church, and in 1928 a school and community hall were built. As the band increased in numbers, interest on the capital was not enough to pay the high per-capita supplements, and the capital fund began to decline sharply. The ever-practical Dokis Indians, seeing no sense in squandering their capital resources, reduced individual payments to $13 per person per month in 1965 and discontinued them altogether in 1976.

In the mid-1950s the band decided to use large amounts of capital to improve the quality of life on the isolated reserve. And so in 1956-57 a road was built from the reserve to connect with Highway 64 at Monetville. In 1958 a power line was constructed, and, from time to time, new houses were erected for band members. In 1972 a modern band office was built, and in 1976 a fire hall and fire truck were added. In March 1986, there was still $124,336 left in the capital fund from the timber sale of 1908.

The band had ceased issuing logging permits in 1956 and had organized its own logging company, which by 1971 was employing twenty-six band members, who were each paid twenty dollars a day, leaving the company with a profit of $60,000. As of 1988 the company was continuing to log on the reserve and on adjacent crown land, leased from the Ontario government. About 1½ million board feet of sawlogs is cut annually and sold to local sawmills. To ensure an income from the forest for future generations, the band reforests the reserve regularly with pine seedlings, which grow extremely well in the acidic soil and favourable climate. Some band members operate tourist lodges and marinas on the reserve, thereby earning good incomes.

Today, the band numbers slightly over 300. About 190 members live on the reserve, the two original family names — Dokis and Restoul — predominating. They are a proud and prosperous people, these Indians who now call themselves "Eagles on the River." Like their Eagle ancestor, Michel, they jealously guard their independence and plan, as they always have, for the future. And like their ancestors, whose sentimentality angered Walton a century ago, they, too, are a sentimental people. Every year they hold a cultural and historical fair to remind themselves of their past. The centre-piece of the village is a carefully preserved log house, one of the originals built by their ancestors when they moved onto the reserve in 1894 to guard their timber from the encroachment of lumber companies. Thus this small band of Indians has mastered the art of combining reverence for the symbols of its continuity as a people with the ability to adapt and innovate in the modern world. Therein lies the secret of its survival and progress.

Chapter Fifteen

The Deserted Village

After its timber was cleared from the Dokis Reserve in the spring of 1920, the Georgian Bay Lumber Co. had no more trees to cut. Consequently, after fifty years of continuous logging the company was faced with a major decision: to look for more timber limits or to close the mill. To continue to saw pine, the only wood the mill had ever sawn, the company would have had to seek timber limits as far away as Sault Ste Marie and beyond. Pine timber limits were then extremely expensive, and the cost of hauling logs the many miles from Sault Ste. Marie to Waubaushene would have been exorbitant. As an alternative, the company could have begun sawing spruce and hemlock, but profits from these woods were small. And so, reluctantly, Sheppard and the other directors decided to close the mill that had operated continuously ever since William Hall had established it sixty years before.

The four to five hundred workers that the company normally employed were laid off, and the sixteen hundred people in the once thriving trading centre of Waubaushene lost their main economic base, just as Port Severn had twenty-four years earlier. In reporting the mill's closure, the *Canada Lumberman* expressed the gloomy sentiments of everyone in the village: "It seems a sad comment on the industrial activities and development of the day that this modernly equipped and ably conducted plant should fall into disuse solely through lack of raw material . . . and that Waubaushene, a hitherto prosperous community . . . should become like 'The Deserted Village' described by Oliver Goldsmith."[1]

With typical foresight, Sheppard had protected himself financially against the inevitable closure of the mill. In September 1917

Parliament had passed An Act to Authorize the Levying of a War Tax upon Certain Incomes[2] — Canada's first income tax act. The tax was progressive, and the rates were quite high. On an income dividend of $40,000, for example, a shareholder in Georgian Bay Lumber would have had to pay a tax of $7,200, or 18 per cent. Capital gains dividends, however, were exempt, and Sheppard used this knowledge to extract the last profits of the company without paying taxes on them.

In the years from 1917 to 1920 the company paid out only small annual dividends. The balance of the profits was converted into bonds, registered in the company's name, or allowed to accumulate in its bank account. By 1920, when it was decided to close the mill, the company was holding well over $1.1 million in cash and bonds and, with other liquid assets, was worth over $1.5 million. The share capital was still fixed at the original $200,000.

On 6 November 1920, the directors passed a by-law increasing the capital stock of the company from $200,000 to $1.5 million, equalling the amount of the company's reserves and undistributed profits since 1917. This was to be done by issuing 13,000 shares of $100 each. These were to be distributed among the shareholders in proportion to the number of original shares each held. The by-law was confirmed by supplementary letters patent issued by the secretary of state on 30 November, but in actual fact only 9,725 new shares were issued, increasing the share capital of the company to $1,172,500.

To get the cash out without paying tax, the directors simply reduced the capital by paying each shareholder a capital dividend. Thus, on 13 May 1921 a by-law was passed "cancelling or extinguishing the unissued share capital and . . . returning to the holders of the 11,725 shares outstanding paid up capital . . . $90 per share, [thus] reducing the amount of each of the outstanding shares from $100 to $10, and the capital to $117,500."[3] The by-law was confirmed and the capital reduction approved by supplementary letter patents issued on 13 July.

Dividends paid in 1921, as a consequence of the capital reduction, amounted to $972,500. Stuart Dodge received $127,261; Josephine and her sons got $394,835. On his 450 shares, Sheppard received a dividend of $218,812.50. The other shareholders received amounts ranging from $12,156.25 to $60,781.25, depending on the number of shares each held. No one paid any income tax. Had Sheppard, for example, received his share as income rather than capital dividend, he would have paid in tax, according to the rates established by an amendment to the Income Tax Act in 1919, 60 per cent, or $131,287.50.

In May 1925 the directors distributed the bulk of the remaining assets, tax exempt, by reducing the share capital to $11,725. This time each shareholder was paid $9 per share as capital dividend, thus reducing the share value to $1.

Those who owned shares in Georgian Bay Lumber had kept them through the years, but there was some redistribution of the stock among families. Distribution of Arthur Dodge's 1,075 shares among his family has already been discussed. Also, as mentioned, W.J. Sheppard bought James Scott's 225 shares, giving him a total of 450. Charlie Sheppard bought 25 shares from his uncle Tom, leaving Tom with 85. When Henry L. Lovering died, he left 175 of his 225 shares to his sons; Henry L. Jr received 125 shares, and his brother, W.J. Lovering, inherited 50. Lovering's other 50 shares were held by the Royal Bank, as were the 25 shares originally owned by Joseph Hartman. W.H.F. Russell and Charlie Stocking kept the 75 and 40 shares each had held, respectively, since 1893.

The administrative staff changed only slightly through the years. Sheppard remained as president and general manager; Marshall Dodge was vice president, and Charlie Sheppard was superintendent until 1916, when he also became secretary treasurer. William Gill managed of the mill, and W.H.F. Russell remained as stores manager, a position he held for forty-four years.

Staff salaries paid by Georgian Bay Lumber through the years had always been reasonable. An income tax return for the year 1917 lists the following incomes: W.J. Sheppard, $5,000; Charlie Sheppard, $4,000; Ella Breech, $1,258; Marshall Dodge, $1,000; William Gill, $1,800; W.H.F. Russell, $2,000; H.L. Lovering Jr., $2,000; Leigh Sheppard, $2,500. Woods managers and other senior employees received from $1,040 to $1,200.

After the share value had been reduced to one dollar, Sheppard decided to gain absolute control of the company by buying the outstanding shares. He managed to acquire all of them but 146. For some reason — perhaps sentiment — Marshall Dodge kept six shares and continued as vice-president. Charlie Sheppard kept his forty shares, and Russell and Stocking, although asked to sell, decided to keep theirs, for it seems that they did not completely trust Sheppard. They knew that he had "sawdust in his blood," and as long as the charter remained active there was always the possibility that he might revive the company and pursue some new lumbering venture somewhere. In that event, they wanted to be in a position to ride on any new wave of prosperity which they knew Sheppard was quite capable of initiating. But that did not happen. By then the mill was gone, having been torn down and the equipment sold to a junk-dealer in 1925. The only assets held by the company then were the land, a few buildings, and some houses in Waubaushene.

Charlie Stocking had not rejoined Georgian Bay Lumber when he returned from overseas in 1918. He joined a lumber company in British Columbia instead. That venture proved unsuccessful, and so Stocking returned to Waubaushene, where he lived in semi-retirement on his investments. In time, he became the "grand old man" of what was left of the village, active in social and cultural activities. Stocking had married Agnes Anderson, Sheppard's private secretary. A woman of Scottish descent, from Ayr, Ontario, Agnes had decided, several years ahead of her time, to pursue a traditional male's occupation as secretary and clerk in a business establishment. She joined Georgian Bay Lumber in the 1890s and stayed with them until her marriage.

Charlie Stocking died in a drowning accident in 1954 at the age of ninety. He and Agnes, who predeceased him in 1946, are buried in the Newmarket Cemetery.

Agnes had been replaced at Georgian Bay Lumber by another competent woman, Ella Breech, daughter of saw filer Morris Breech, who had come to Canada with Anson Dodge in 1870. Breech married Rachel Gill, and so Ella was related to Sheppard's children. Ella was raised at Port Severn, where her father was head filer until the mill burned in 1896. The family then moved to Waubaushene where Ella, at the age of eighteen, began teaching school. Because she did not enjoy teaching ninety children crowded in one room, she quit and joined Georgian Bay Lumber, first as a stenographer and later as head bookkeeper when Agnes Stocking left. She remained with Sheppard until his death and afterward served as secretary-manager for Charlie Sheppard. Ella died in 1957.

Several other young local women were sometimes employed as secretaries and stenographers by Georgian Bay Lumber. Frances Bettes and Ella's sister, Grace Breech Miller served in the final years of the company's existence.

Waubaushene did not become quite the deserted village the *Canada Lumberman* predicted, but the village's golden era was definitely over when the mill was closed. Before 1920 the village boasted

Charles H. Sheppard 1876-1967   – courtesy Margaret Bell

Ella Breech 1878-1957   – courtesy Rissah Mundy

Burner at Waubaushene Mill being taken down, July 28, 1925
– courtesy Frances Wellman

eleven stores, including a tailor shop, millinery, bakeshop, butcher shop, and several livery stables. Most of these were forced to close or were sold as the population inevitably declined.

Many of the mill hands and unskilled workers moved away. Older workers retired, and those who decided to stay in Waubaushene found other means of supporting themselves. Charlie Sheppard retired to a farm that he bought on the outskits of Aurora, Ontario. Some young men found employment with the Hydro Electric Power Commission (now Ontario Hydro) as linemen and power-house operators; others worked for the railways, and still others found seasonal employment as sailors with Great Lakes steamship companies.

The burgeoning tourist trade provided another source of income. Some of the larger Georgian Bay Lumber Company houses were bought and converted into summer hotels. These did a thriving business with fishers and fall duck hunters, especially after improved roads and Henry Ford's cheap automobile made the village readily accessible to those who sought such diversions.

In 1922 Sheppard had the former Dodge compound and the lumber yard surveyed into town lots. The company then began selling its property. Vacant lots sold for from $30 to $150, depending on size and location. Lots with houses sold from $150 to $750. Mrs W.H.F. Russell bought the family home on Willow Street and a piece of the former Dodge property for $1,500. Russell bought the main store and surrounding property for $2,800. He and his sons operated the store until his death in 1933, when Charles and Clarence took over the business. An older brother, Frank, bought the company's former office at Port Severn and converted it into a store and post office, which he operated for over forty years. The Women's Institute bought the library-reading room for $175.

Stocking bought several lots on the hill which had been part of the Dodge compound but now formed part of upper Elm Street. In 1923 the Stockings built a fine home on the hill and moved out of the company house on Willow Street which they had occupied since their marriage. Stocking also bought some property at the water's edge, at the bottom of Elm Street where he erected a boat-house.

In 1925 the once-magnificent Second Empire summer home of Arthur Dodge, the company's boarding-house since 1913, was sold. A real estate agent named Middleton, from Midland, bought the house and ten surrounding lots for $2,400. The same day he transferred the house and property to John and Elizabeth Palmer, farmers from Thorah township, for $3,200. The Palmers converted the house into a hotel — the Palmer House — which they operated for five years. In 1930 they sold the hotel to Elizabeth Hisey of Midland for $3,500. The next year it was completely destroyed by fire.

Most of the company's property had been sold by 1925, the proceeds having been placed in the profit and loss account of the company and distributed to shareholders with the capital reduction that year. Then, apart from one or two small lots that remained unsold, the only property owned by the company was the office. The record shows that the last annual meeting of shareholders was held on 13 April 1927.

From then on Georgian Bay Lumber was a company in name only and Sheppard, holding most of the by then almost valueless stock, was, for all intents and purposes, sole owner. He transferred $18,000 of Famous Players bonds from his personal portfolio into the company's possession. Interest from these and incomes from small mortgages given for some of the recently sold property paid for maintenance of the office and Ella Breech's salary. Georgian Bay Lumber stationery and envelopes were kept and used for Sheppard's personal and business correspondence.

Like a retired general who keeps his rank, Sheppard maintained the title of president. Following the practice of thirty-five years, he travelled by train to Waubaushene every Monday and spent the week there, living in an apartment that he created on the second floor of the office building. Frequently, he travelled to Toronto to attend boards of directors' meetings of various companies to which he had been appointed.

Sheppard's reluctance to surrender the company's now-useless charter and to abandon the routines of nearly half a century confirm his innate conservatism. But they reveal more than that, for this tough-minded, achievement-driven entrepreneur, like the practical-minded Dokis Indians with whom he sparred for so many years over timber rights, was also a sentimentalist. And so for several more years the company was kept alive, and its president took his place in the office every Monday morning.

One might wonder what an old man found to do all week long in that half-deserted village, peopled with so many ghosts, or in the office so crammed with memories and memorabilia. He probably spent some of his time dreaming the dreams old men are supposed to dream, but Sheppard never lost the young man's ability to see visions. Much of his time was spent converting his visions into

Waubaushene c1900. Left to right: Union Church, built 1881;
two storey school, built 1889; one room school, built 1880
– courtesy William Russell

Final meeting of I.O.D.E., Waubaushene, July, 1928
– courtesy C.A. Stocking

dollars or rearranging and refinancing investments that had resulted from earlier visions.

Sheppard had always been a visionary. Even as he was expanding the timber reserves of Georgian Bay Lumber, he saw clearly that Georgian Bay timber was doomed. In 1899 he began investing in lumbering operations in British Columbia, where an assault on the forests was just beginning. He subsequently became president of the West Coast Lumber Co.

When the sulphate-pulping process for converting wood fibres into paper was invented, Sheppard was one of the first to see a future for the pulp and paper industry in Ontario, where spruce outnumbered pine. In association with J.R. Barber, fabled paper king of Canada, Sheppard invested in the installation of the first ground wood plant and in the erection of the first paper-making machine to use the sulphate-pulping process in Ontario. These facilities belonged to the Spanish River Pulp and Paper Co. at Espanola, of which Sheppard became vice-president. Later, because of his interest in and knowledge of paper-making, he was made a director of the Toronto Paper Co.

Sheppard had the ability to spot opportunities for investment and usually took advantage of them, as the following incident illustrates. On his many trips to attend directors' meetings in Toronto, he frequently invited one or more of his grandchildren to have luncheon with him. Once, in 1924, he noticed that his granddaughter, Nora Gray, a student at the University of Toronto, was wearing a pair of the new-fashioned rayon stockings. Viscose rayon, a byproduct of the pulping process, had just made its debut as a textile fabric, and Sheppard was curious about it. "Do you like those stockings?" he asked Nora. "Do the other girls wear them? How do they stand up to wear?" he queried. It seems he had detected yet another investment possibility on the shapely legs of his young granddaughter.

When gold was discovered in Kelly township, near Lake Wanapitei, in 1896, precipitating a stampede of prospectors into the area, Sheppard typically had been far ahead of the pack. Gold deposits had been found on the company's timber limits as early as 1892. Sheppard formed the Wannipitae Mining Co. with W.H.F. Russell, James Scott, Oliver Barton Sheppard, Tom Sheppard, Peter Wallace, and a prospector named Thomas Thomas. The men acquired eleven mining licence — among the first to be issued in Ontario — covering about 1,212 acres in the townships of Neelon and Dryden, just south of Ramsay Lake, within the present boundaries of Sudbury. There was not enough gold to justify a mine, and the licences were allowed to expire in 1900, but Sheppard's interest in mining remained high.

He eventually struck pay-dirt when he and his partner, J.P. Bickell, invested in Sandy McIntyre's gold claims near Porcupine, Ontario. The resulting McIntyre Porcupine gold mine became one of the legendary success stories in Canadian mining, making Sheppard a very wealthy man. Some of the other mining stocks he invested in turned out to be "dogs." The profitable ones, apart from McIntyre Porcupine, were Castle-Tretheway Mines, of which Sheppard became vice-president, the International Nickel Co., Sudbury Basin, and Hollinger Gold Mines.

In the mid-1920s, Leigh Sheppard bought an airplane. He dropped, unannounced, into Georgian Bay one weekend and taxied up to the dock at his father's cottage, near Port Severn, to show off his new toy and to give rides to any of his relatives daring enough to go up. The first to volunteer was his ageing father. From the vantage point of 5,000 feet in the air, Sheppard saw more than the contours of the land and water he had tramped and boated over nearly all his life; he saw the potential of the flying contraption in the mining business. Soon afterward he and Bickell leased an airplane and hired a pilot, for mining exploration.

Sheppard quite accurately predicted that the growing motion-picture craze of the 1920s would be a lucrative field of investment. Although he was not one to idle away his own precious time in movie houses, he saw thousands of others doing so. He invested heavily in the entertainment business. He bought bonds ($266,000) in Famous Players Canadian Corp.; he invested in the St. Denis Theatre in Montreal ($25,000), Paramount Public Corp. of New York ($5,000), Eastern Theatres Ltd ($36,500), Goldwyn Pictures ($5,000), Hamilton United Theatres ($22,500), and Marcus Loew's Theatres Ltd of Toronto ($27,100).

One thing Sheppard did not predict was the stock market crash in 1929. Even so, he was not as badly injured as many others. Shrewd enough not to have put all his investment eggs in one basket, he had about 50 per cent of his pre-crash assets ($2,616,000) in bonds. Like many others, he suffered great losses in stocks, real estate, and some large loans to developers. Stocks with a pre-1929 purchase value of about $1,712,000 had a market value of only $775,094 in 1935. But because his investment portfolio consisted of a judicious balance among banks, grocery stores, manufacturing, mining, real estate, theatres, and transportation stocks, his losses were minimized.

The stock market crash and time began to take their toll on Sheppard. By 1930 his health had begun to deteriorate; consequently, he spent less time in Waubaushene and more in Miami, Florida, where he was in the habit of passing the winter months. It was in Miami, on 3 November 1934, while fully dressed and engaged in conversation with members of his family, that he died. He was in his eighty-third year. His body was returned to Canada and buried in the family plot in the Coldwater Cemetery.

The young pastor of the Coldwater United Church who conducted Sheppard's funeral service seems to have had some difficulty in choosing an appropriate passage of scripture with which to eulogize the enigmatic Sheppard. In the end, he chose, with remarkable insight, a little known verse from Psalm 90: "For all our days are passed away in thy wrath; we spend our years as a tale that is told." Though it is doubtful if Sheppard ever feared the wrath of God or anyone else, his life, as the pastor clearly saw, was one with the Georgian Bay Lumber Co. with which he had been identified for so many years.

Sheppard was never as wealthy as his acquaintances and many of his critics supposed, but he was worth a lot of money. In the saturnian decade of the 1920s, his investments may have reached $10 million — some claimed they were worth $15 million — but when he died, in the depth of the Depression, his estate was assessed at a mere $3.7 million. It was all left to his family. Each grandchild received $1,000. This represented less than one-half of one per cent of the estate, but he reckoned it enough to confirm his affection for them, while ensuring that none would have the unfortunate wherewithal "to start at the top." He left his housekeeper, Maud Skene,[4] $10,000 in cash, and the rest was divided equally among his five sons and one daughter. Churches,[5] charities, and cultural institutions received nothing, but Sheppard's family did use some of their inheritance to build a library in Coldwater in memory of their parents.

The tale of Georgian Bay Lumber did not end with Sheppard's death. When his will was finally probated in 1936, his son, Charlie, acquired ownership of the company, and like his father he kept it alive, partly for reasons of the heart, but mainly as a locale for summer vacations. As his father had done for ten years, Charlie maintained the office as a base from which to manage his own small lumber company and other financial interests with which he was involved.

It seems that Charlie did not adjust easily to the role of gentleman farmer. The farm at Aurora had really been purchased for the benefit of his son, Reg, a student at the Royal Agricultural College in Guelph, not because of any strong interest in farming on Charlie's part. By 1933, Reg had taken over management of the farm, and Charlie, who had never been able completely to purge his system of lumbering — the only occupation he had ever known — came out of retirement.

Charlie and his friend and former partner, Jack Dunn, reincorporated the Sheppard and Dunn Lumber Co., which had been woundup in 1924 along with all the other Sheppard lumber companies. The headquarters of the new company, like that of the former one, was at Waubaushene. Jack Dunn was president and held half of the 100 shares. Charlie held 48 shares, his brother Leigh held one share, and the ever-faithful Ella Breech received one share, was made a director, and served as secretary treasurer. The firm's lumber was sawn in the mill of the McFadden Lumber Co. at Blind River and sold through the outlets of the Sheppard and Gill Lumber Co. of Toronto. It was not a big outfit, having share capital of only $10,000, but it was large enough to occupy Charlie's interest.

Charlie Sheppard's main interests, apart from lumbering, were big, fast cars and boats. His favourite automobile was the Packard. In 1934, he bought a large yacht, *Ambler*, from the owners of the Leggett Drug Co. of Boston. Originally called *Cynthia*, *Ambler* was a truly magnificent craft and, like *Skylark* in the nineteenth century, was the pride of Waubaushene, where she was based. Built in 1922, she was 129 feet long and had a 23-foot beam, with a steel hull. Her twin screws were powered by two 250-horsepower Winston diesel engines. Charlie and his family used the *Ambler* as a summer home at Waubaushene and for pleasure cruising on the Great Lakes.

When the Second World War started, Charlie could see no sense in trying to maintain *Ambler*; consequently, he gave her to the Royal Canadian Navy. She, like several other yachts donated by prominent Canadians, was refitted and used as a patrol vessel on the Atlantic coast. Known officially as Z32, she was the only Canadian patrol vessel that did not carry a gun.

Under Charlie's ownership, Georgian Bay Lumber continued to be a corporation in name only. Interest from the Famous Players bonds, still registered in the company's name, paid the bills. To comply with the requirements of the secretary of state, annual returns were filed. Probably out of respect for his father's memory, Charlie never adopted the title of president. He is listed in the annual statement of affairs as vice-president; Ella Breech was secretary treasurer.

For the next fifteen years, Charlie spent a good deal of time at Waubaushene, especially in the summer months. He slept in the facility created by his father in the second storey of the office and ate in the company's former machine shop, which had been converted into a kitchen and dining place. He was as penny-pinching as his father, perhaps even more so. Probably for this reason, he decided to stop paying the five dollar fee for filing annual returns to the secretary of state. And so the company's books were closed on 9 August 1942, and an application for surrender of the charter[6] was submitted to the secretary of state. The surrender was accepted in October, and the Georgian Bay Lumber Co. ceased to exist. The firm — which had once been capitalized at $1 million, which in its sixty-one years had produced billions of board feet of pine lumber, and which had earned millions of dollars for its various owners — had a closing book balance of $19,326.89, of which $18,000 was Sheppard's Famous Players bonds.

Charlie transferred ownership of the office and property into his own name. He continued to spend some time in Waubaushene, managing the Sheppard and Dunn Lumber Co. until 1956, when he sold the office to the road contractor who built part of Highway 69. The contractor used the building as an office and store-house until the highway was finished, at which time the building was abandoned and the property reverted to Tay township. For the next fifteen or more years, the once-busy office stood empty in a grove of Waubaushene's ubiquitous willow trees, its paint peeled, its windows smashed. Finally, in 1977, young boys, ignorant of and perhaps uncaring for Waubaushene's traditions and the office's history, made a bonfire out of it.

The Sheppard and Dunn Lumber Co. was sold to the Dominion Chemical Co. in 1965. Charlie Sheppard died in 1967, the last surviving shareholder of the legendary lumber company that had contributed so much to the early development of the Georgian Bay district. He and his wife, Ellen, who predeceased him in 1961, are buried in the Newmarket Cemetery.

Little remains in the still-picturesque village of Waubaushene to remind one of the existence of the Georgian Bay Lumber Co. Some of the original houses still stand and are occupied, but every year one or more of them falls victim to fire or reconstruction. The brick machine shop, happily immune to the arson's match, is now owned by the township and used as a warehouse. The library-reading room, which stood empty for many years was recently purchased, moved, and restored. Today it serves as a showroom for new and restored wicker furniture manufactured by Mrs. Pam King, one of Waubaushene's new entrepreneurs. The Odd Fellows' hall, formerly a school, originally a Catholic church, is used regularly by the IOOF for meetings and socials.

Only the Union Church exists in its original state. Beautifully maintained, it provides a place of worship for the Anglican and United Church congregations, as it has every Sunday since its dedication in 1881. William Earl and Arthur Dodge would be pleased. And perhaps Anson Dodge might argue that the church's survival after all fulfills his prophecy "a Deo Victoria."

NOTES

CHAPTER ONE

[1] Bracebridge *Northern Advocate* (undated), in Clippings from Muskoka Papers, John P. Robarts Library, University of Toronto.

[2] Toronto *Mail*, 16 August 1872.

[3] Newmarket *Era*, 5 July 1872.

[4] Orillia *Northern Light*, 3 July 1872.

[5] National Archives and Records Service, Suitland, Maryland (hereafter NARS), RG 21, US District Court for the Southern District of New York, Bankruptcy No. 4336.

[6] Hamilton *Times*, quoted in Newmarket *Era*, 16 May 1873.

[7] Ibid.

[8] Muskoka Papers.

[9] Toronto *Mail*, 19 June 1872.

[10] Toronto *Globe*, 15 June 1872.

[11] NARS, RG 21, Bankruptcy No. 4336.

[12] Daniel James to William Earl Dodge, April 1859, in Phyllis Dodge, *Tales*, 148.

[13] NARS, RG 21, Bankruptcy No. 4336.

[14] Ibid.

[15] Ibid.

[16] Ibid.

CHAPTER TWO

[1] Simcoe County Archives, Minesing, Ontario (hereafter SCA), The Georgian Bay Lumber Company Papers, Maganettewan Lumber Company Inventory, 1 May 1871.

[2] *Directory of Simcoe County, 1972*, 275.

[3] SCA, Georgian Bay Lumber Company Papers, Maganettewan, 1 May 1871.

[4] The Magnetawan River had many spellings. The one the company used seems to have been the original. Even Dodge had trouble with the name. He once told a bankruptcy court that the river was called the "Maganalan."

[5] Canada *Gazette*, July 1870 to June 1871, 874.

[6] Before passage of the Ontario Joint Stock Letters Patent Act in 1874, Ontario joint stock companies could be formed only by special acts of the legislature.

[7] Ontario, *Statutes*, 1869, 34 Vic., c. 71, An Act to Incorporate The Georgian Bay Lumber Association.

[8] Ontario, *Statutes*, 1872, 35 Vic., c. 100, An Act to Incorporate the Maganettewan Lumber Company of Ontario.

[9] David Gibson came to Upper Canada in 1825 from Scotland, where he had qualified as a surveyor and engineer. A Reformer in politics, he took part in the rebellion of 1837, subsequently fleeing to the United States. He returned to Canada in 1848. In 1853, he was made inspector of crown lands agencies and superintendent of colonization roads, a position he held until his death in 1864.

[10] Ontario, *Statutes*, 1872, 35 Vic., c. 98, An Act to Incorporate the Parry Sound Lumber Company.

[11] Hamilton, in *East Georgian Bay Historical Journal*, 1983, 38.

[12] Ibid.

[13] It is not known when the timber mill was built at Muskoka Mills, but it was probably in the 1860s. It seems to have been unique in the Georgian Bay area at that time. Timber was generally squared in the bush, hauled to streams and rivers, and floated out to Georgian Bay for rafting to the railhead at Collingwood. Timber and sawlogs floated down the turbulent Muskoka and Musquash rivers were severely damaged in the falls and rapids. Because bruised and battered timber fetched a low price on overseas markets, the operators of the Muskoka Mills found it advantageous to float logs down the rivers and square them with a timber saw at the mouth of the river. Other lumbermen operating in Muskoka and also using the Musquash River for running timber were not so fortunate. In 1874 the Cook Brothers, who operated mills in Midland, petitioned the attorney general for slides to be built to facilitate the descent of timber and sawlogs, similar to those operated by the province on the Gull and Burnt rivers for lumbermen in Haliburton. The province rejected the request but passed legislation in 1881 permitting joint stock companies to be formed, for the purpose of improving streams and rivers for the descent of timber and with authority to charge tolls to other users. Under this legislation, the Muskoka Slide and Boom Co. was formed in 1882, and slides were constructed on the Muskoka and Musquash rivers.

[14] *Indian Treaties and Surrenders*, Treaty No. 48, 117.

[15] National Archives of Canada (hereafter NAC), RG 10, vol.128, Rama, Snake Island, and Coldwater Indians to Sir Charles Bagot, 26 May 1842.

[16] Simcoe County Registry Office (Barrie), Tay Township, concession 13, lot 19, 10751.

[17] The night after Christie's mill burned, someone tried to burn George Caswell's grist mill in Coldwater. A piece of smouldering wood had been pushed through an opening in the wall, but it was discovered by a young boy before any great damage was done. This mill, like the the Port Severn sawmill, had been built by the colonial government for Indians in 1830. George Caswell bought the mill in 1836, when the Indians were moved off the good farm land around Coldwater to new reserves at Rama, on Lake Couchiching, and at Beausoleil Island, in Georgian Bay.

[18] Gibbard's map of 1853 identifies the mill as " Thompson's Mill," showing it on lot no.6 near the boundary with lot no.7.

CHAPTER THREE

[1] Aubrey White was employed by the Crown Lands Department as a woods ranger (1876-8) and later as crown lands agent at Bracebridge. In 1882 he joined headquarters staff in Toronto. A cool, capable administrator schooled in the tradition of the British civil service, White rose through the ranks, eventually becoming assistant commissioner of crown lands, and in 1905 he became the first deputy minister of the Department of Lands, Forests and Mines. He died in 1915.

[2] William Earl Dodge sold *Mittie Grew* to the Parry Sound Lumber Co. in 1876.

[3] Ontario, *Statutes*, 1869, 32 Vic., c. 30, An Act to Incorporate the Toronto, Simcoe and Muskoka Junction Railway Company.

[4] Bracebridge *Northern Advocate*, 17 February 1871.

[5] Ontario, *Statutes*, 1873, 36 Vic., c. 74, An Act to Incorporate The Lake Simcoe Junction Railway Company.

[6] NARS, RG 21, Bankruptcy No. 4336.

CHAPTER FOUR

[1] Canada *Gazette*, July 1870 to June 1871, 874.

[2] Ontario *Statutes*, 1872, 35 Vic., c. 99, An Act to Incorporate the Georgian Bay Lumber Company of Ontario.

[3] It is not clear where the church stood. In 1882, it was torn down and the property transferred back to the Georgian Bay Lumber Co. In return, the company gave the diocese another piece of property, approximately one-fifth of an acre in area. A new church was built in 1883. In 1915, the diocese sold the new church to the Public School Board, S.S. No. 12, Tay, for

$1,000. In 1925, after the mill had closed and the population of Waubaushene had decreased, the school was sold to the Independent Order of Odd Fellows. It has been remodelled, and an addition put on, but it still has a church-like appearance and is probably one of the oldest buildings in Waubaushene.

[4] Cleland, *Phelps Dodge*, 22.

[5] Stuart Dodge, *Memorials*, 36.

[6] Phelps, Diary, November 1836, in Phyllis Dodge, *Tales*, 73.

[7] Ibid.

[8] Bracebridge, *Northern Advocate*, no date.

[9] Ibid.

[10] Ibid.

[11] Bracebridge, *Northern Advocate* no date, in Muskoka Papers.

[12] Kirkwood and Murray, *Undeveloped Lands*, 76.

[13] Orillia *Packet*, 23 November 1873.

[14] Ibid.

[15] Ibid.

CHAPTER FIVE

[1] McMurray, *Free Grant Lands*, 67.

[2] Orillia *Northern Light*, 10 November 1871.

[3] The figures quoted in this chapter may be found in the Georgian Bay Lumber Co. file in the Simcoe County Archives at Minesing, Ontario.

[4] The original mark of the Georgian Bay Lumber Co. was a large G with small numbers, representing camp numbers, beside it. When William Earl Dodge took over the company in 1874, the mark was changed to WED. Perhaps out of respect to William Earl, the same mark was used for forty years after his death, for as long as the company produced lumber.

[5] *Canada Lumberman*, 1 August 1881.

[6] For a complete description of timber-making in Canada in the nineteenth century see Whitton, *A Hundred Years*, 114-30, or MacKay, *The Lumberjacks*, 77-78.

[7] Waney timber was octagonal-shaped rather than square. It began appearing in 1861, its purpose being to reduce the amount of wood wasted when logs were squared.

[8] Ontario, *Sessional Papers, 1880*, Paper no. 4, Report of the Commissioner of Crown Lands of the Province of Ontario for the Year 1879.

[9] Ibid.

CHAPTER SIX

[1] NAC, Macdonald Papers, MG 26 A, vol. 345, 158,237, A.G.P. Dodge to John A. Macdonald, 24 July 1873.

[2] Canada, *Statutes,* 1872, 35 Vic., C. 118, An Act to Naturalize Anson Greene Phelps Dodge.

[3] Bracebridge, *Northern Advocate*, 17 February 1871.

[4] The Conservative party established the *Mail* in 1872 to counteract the editorial bias of George Brown's *Globe*. Dodge subscribed for stock in the *Mail*.

[5] Toronto *Mail*, 19 June 1872.

[6] NAC, Macdonald Papers, John A. Macdonald to Alfred Boultbee, 18 June 1872.

[7] Ibid., Macdonald to Angus Morrison, 18 June 1872.

[8] Ibid., Macdonald to John B. Robinson, 18 June 1872.

[9] Newmarket *Era*, 21 June 1872.

[10] Originally called the Toronto and Georgian Bay Canal, the Huron and Ontario Ship Canal was designed to connect Toronto and Georgian Bay by following the Humber and Holland rivers into Lake Simcoe and going from Kempenfeldt Bay to Georgian Bay by way of Willow Creek and Nottawasaga River. The main objection to the canal was the large amount of excavation that would be necessary to cross the Oak Ridges moraine south of Newmarket and, in effect, to reverse the flow of the Holland River from Lake Simcoe into the Humber. Further, nine miles of very expensive excavation would be required to connect Kempenfeldt

Bay with Willow Creek. The canal had much appeal to the residents of York North, because it would pass through the centre of the county. The Reformers had supported the canal since its charter was incorporated in the 1850s.

[11] Orillia *Northern Light*, 5 July 1872.

[12] Newmarket *Era*, 21 June 1872.

[13] Toronto *Globe*, 24 June 1872.

[14] Ibid., 20 June 1872.

[15] Newmarket *Era*, 5 July 1872.

[16] Ibid.

[17] Patstone, *Saint Paul's Church*.

[18] William Earl Dodge Jr to William Earl Dodge, 1850, in Phyllis Dodge, *Tales*, 147.

[19] Anglican Church hearing into the Ramsay affair, reported in Toront *Globe*, 15 April 1873.

[20] Ibid.

[21] A.G.P. Dodge to George Brown, 17 March 1873, reproduced in the Newmarket *Era*, 28 March 1873.

[22] Toronto *Globe*, 24 March 1873.

[23] Ibid.

[24] The *Courier* was a Conservative newspaper, published in Newmarket.

[25] A.G.P. Dodge to George Brown, 17 March 1873, published in Newmarket *Era* 28 March 1873.

[26] Ibid.

[27] Canon Ramsay to the editor, Newmarket *Era*, 21 March 1873.

[28] Ibid.

[29] Statement of Dodge in the House of Commons, 17 March 1873, reported in Toronto *Globe*, 18 March 1873.

[30] Toronto *Globe*, 14 March 1873.

[31] A.G.P. Dodge to Brown, Newmarket *Era*, 21 March 1873.

[32] This case involved Dr Robert Ramsay and Dodge's agent, Charles Ostrander. Ostrander, who suffered from severe depression and acute alcoholism, was a patient of Dr Ramsay's. When it became apparent that Ostrander might take his own life, Ramsay allegedly insured him for $15,000 and sent him off to California, where Ostrander committed suicide.

[33] Newmarket *Era*, 7 June 1873.

[34] Ibid., 18 July 1873.

[35] Ibid.

CHAPTER SEVEN

[1] William Earl Dodge Jr succeeded his father as president of Phelps Dodge in 1880.

[2] NARS, RG21, Bankruptcy No. 4336.

[3] Dodge, *Memorials*, 38.

[4] *Monetary Times*, 30 May, 1873.

[5] NARS, RG 2l, Bankruptcy No. 4336.

[6] Ibid.

[7] As an example of how extravagantly Anson Dodge lived, he rented a house on East 34th Street from William B. Astor, son of nineteenth-century fur magnate John Jacob Astor, for the princely sum of $4,800 a year. Only Rebecca and A.G.P. Dodge Jr lived in the rented mansion after 1869, when Anson spent most of his time in Canada and had, in fact, built his own luxurious home at Roche's Point.

[8] NARS, RG 21, Bankruptcy No. 4336.

[9] Ibid.

CHAPTER EIGHT

[1] NARS, RG 21, Bankruptcy No. 4336.

[2] NAC, Macdonald Papers, reel C 1756, 179,903, A.G.P. Dodge to Sir John A. Macdonald, 28 April 1882.

[3] Percival Dodge to Eugenia Price, 27 June 1963.
[4] NAC, Macdonald Papers, reel c 1759, 184,111, W. J. Macaulay to Macdonald, 2 November 1882.
[5] Ibid.
[6] Information provided by Mrs Cleveland E. Dodge Jr.
[7] *Past and Present in Vermilion County*, 981.
[8] Jones, *History of Vermilion County*, 9.
[9] Ibid.
[10] Ibid.
[11] Danville *Commercial News*, 28 May 1918.
[12] Jones, *History*.
[13] Danville *Commercial News*, 28 May 1918.
[14] *Past and Present*, 981.
[15] Danville *Commercial News*, 30 May 1918.

CHAPTER NINE

[1] Archives of Ontario (hereafter AO), Georgian Bay Lumber Company Papers (un-catalogued), William Earl Dodge to D'Alton McCarthy, 16 August 1873.
[2] Ibid.
[3] Ibid.
[4] In 1885, Christie sold the Sturgeon Bay property to John S. Playfair of Toronto for $60,000. In 1887, Playfair transferred the property to his son James, who sold it to Edward McLaughlin of Sturgeon Bay in 1897. Subsequently the lots were subdivided and sold to individuals. The mill was closed, probably by James Playfair in the 1890s.
[5] When Melissa Dodge died in 1903, her estate papers showed a debt of $244,000, charged against the estate of Arthur Murray, who had died seven years earlier.
[6] The Georgia Land and Lumber Co. operated for some years under the legal fiction that it was George Dodge's personal property. But bad feelings developed, some blood was shed, and long, involved legal proceedings extended well into the twentieth century. Not until 1923 was the question of title finally settled by the US Supreme Court. During the long period of litigation, the Georgia Land and Lumber Co., managed by the heirs of William Earl Dodge, continued to operate.
[7] On 17 January 1885, the mortgage being fully paid, the executors of William Earl's estate discharged it.
[8] *Canada Lumberman*, 1 June 1881.
[9] Orillia *Times*, 22 May 1884.
[10] *Canada Lumberman*, 30 October 1880.
[11] NAC, George Brown Papers, MG 24 B40, vol. 11, William Earl Dodge to George Brown, 21 September 1877.
[12] Ibid.
[13] Ontario, *Sessional Papers, 1880*, Paper No. 4, Report of the Commissioner of Crown Lands of the Province of Ontario for the Year 1879.
[14] Ontario, *Statutes, 1881*, 44 Vic., c. 11, An Act for Protecting the Public Interest in Rivers, Streams and Creeks.
[15] Ibid., 44 Vic., c. 19, An Act for the Incorporation by Letters Patent and Regulation of Timber Slide Companies.

CHAPTER TEN

[1] Dialogue was reconstructed from discussions recorded in board minutes of Georgian Bay Consolidated Lumber.
[2] In partnership with Titus B. Meigs, Arthur invested in timber land in Oregon at this time. The timber purchased proved to be diseased. Meigs backed out of his commitment, but Arthur paid in full for the useless investment.

[3] AO, Georgian Bay Lumber Co. Papers (Uncatalogued), Minutes of the meeting of the board of directors of Georgian Bay Consolidated Lumber, 14 January 1887.

[4] The production of the five millls in 1886 was as follows: Waubaushene, 26 million feet; Port Severn, 14 million feet; Byng Inlet, 7 million feet; and Collingwood 5 million feet.

[5] The company continued to produce lath, but only from edgings.

[6] In the year 1882, the Midland Railway transported 104,461,000 board feet of lumber, 739,000 cubic feet of square-timber, 63,318,000 shingles, and 698 carloads of fence posts and railway ties, much of it produced in the Georgian Bay region.

[7] AO, Georgian Bay Lumber Company Papers (Uncatalogued), Minutes of the annual meeting of the board of directors of Georgian Bay Consolidated Lumber, 3 February 1888.

[8] Ibid., Minutes, 10 February 1887.

[9] Ibid, Minutes, 3 February 1888.

[10] Ibid.

[11] NAC, RG 95, vol. 2, Petition for Incorporation by Letters Patent of The Dominion Mercantile Company Limited, 22 June 1888.

[12] AO, Ibid, Minutes, 3 February 1888..

[13] Ibid., 15 April 1889.

CHAPTER ELEVEN

[1] Dunnan and Dodge, *Jewells and Dodges*, 9.

[2] Ibid,

[3] New York *Tribune*, 18 October 1896.

[4] Yale, *Obituary Record*, 1897, 476.

[5] Orillia *Weekly Times*, 29 October 1896.

[6] Ibid.

[7] Percival Dodge to Eugenia Price, 27 June 1963.

[8] Ibid.

[9] Ibid.

[10] Anecdote provided by Mrs Margaret Bell, Terra Cotta, Ontario.

[11] Orillia *Times*, 22 May 1884.

[12] The dinner set has been handed down through successive generations of Jesse Peckham's relatives. It is now in the possession of Margaret Bell.

[13] *Coldwater Tribune and Waubaushene Investigator*, 21 March 1889.

[14] Orillia *Times and Expositor*, 12 August 1886.

[15] Newmarket *Era*, 13 August 1886.

[16] Minutes, Annual Meeting, Georgian Bay Consolidated Lumber Co., 10 February 1887.

[17] Orillia *Times and Expositor*, 12 August 1886.

[18] *Coldwater Tribune and Waubaushene Investigator*, 27 June 1889.

[19] Ibid., 6 June 1889.

[20] Ibid.

[21] Ibid., 27 June 1889.

[22] Orillia *Times and Expositor*, 12 July 1888.

[23] *Coldwater Tribune and Waubaushene Investigator*, 4 July 1889.

[24] Ibid., 30 June 1887.

[25] Ibid., 19 June 1889.

[26] Ibid., 17 April 1890.

[27] Orillia *Packet*, quoted in Barrie *Northern Advance*, 3 July 1890.

[28] Orillia *Times and Expositor for East Simcoe and North Ontario*, 28 June 1888.

[29] *Coldwater Tribune and Waubaushene Investigator*, 11 July 1889.

[30] Ibid., 8 August 1889.

[31] Ibid.

[32] AO, RG 1, file 15769/83, James Scott to T.B. Pardee, 29 April 1885.

CHAPTER TWELVE

[1] Merrill, Ring and Co. sold the Maganettewan property to the Holland and Emery Lumber Co. in 1899. Holland and Emery, in turn, sold it to Holland and Graves, later Graves, Bigwood and Co. The Page mill was torn down and a new enlarged mill erected on the mainland, a little east of the present village of Byng Inlet. The mill operated until 1915, when it was closed down for want of timber.

[2] AO, RG 8, file 3579, James Scott to the Provincial Secretary, 15 March 1901.

[3] *Canada Lumberman and Woodworker*, 1 February 1921.

[4] NAC, RG 95, vol. 2, James Scott to the Undersecretary of State, 2 May 1893.

[5] Ibid., Dominion Order-in-Council No. 1395, 13 May 1893.

[6] *Canada Lumberman*, 1 April 1886.

CHAPTER THIRTEEN

[1] *Canada Lumberman*, 1 January 1921.

[2] Orillia *Packet and Times*, 20 March 1924.

[3] Ibid.

[4] Ibid.

[5] *Canada Lumberman*, 1 January 1921.

[6] Orillia *Packet*, 8 November 1934.

[7] Scott left an estate valued at $250,000, a large portion of which was bequeathed to charitable organizations.

[8] Orillia *Packet and Times*, 20 March 1924.

[9] Ontario, *Statutes, 1898*, 61 Vic., c. 9, An Act Respecting the Manufacture of Pine cut on the Crown Domain.

[10] Not all Michigan lumbermen supported the Dingley-Alger tariff. One who opposed it was F.W. Gilchrist, a wealthy lumberman in Michigan, mayor of Alpena, owner of valuable timber limits in Canada, and owner of a Great Lakes shipping and towing fleet. Associated with Gilchrist in Canadian logging were fellow townsmen Albert Pack and G. Fletcher. All three had to close their Michigan sawmills as a consequence of the ban on export of Ontario sawlogs. After the Spanish-American war, a delegation of Alpena citizens approached Gilchrist for a donation toward a monument to commemorate Gen. Alger's services as secretary of war. Gilchrist refused to contribute. Referring to the three idle mills, he said to the delegation, "There are Alpena's monuments to Alger."(see Kauffman, *Logging Days*, 102.)

[11] The *Canada Lumberman*, March 1902, indicated that American mills built on the north shore, as a direct result of the ban on the export of logs, had a total annual capacity of 265 million board feet. While this production provided employment in Ontario and contributed to the provincial treasury, it also accelerated the clearing of pine from the forests of central Ontario.

[12] In 1900, Toronto car lot prices for no. 1 grade, one-inch pine were $32 to $34 per thousand feet; in 1911 they were $60 to $65 per thousand. Prices in Buffalo and Tonawanda were about 50 per cent higher.

[13] In 1900 the total value of forest products exported to the United States was $12,190,617; in 1911, $28,785,427.

[14] Extensive logging operations had been conducted on some of these berths by Georgian Bay Consolidated Lumber, prior to the sale to Merrill and Ring. Scadding and Maclennan townships were reached by a twelve-mile tote road cut through the bush from Wahnapitae Station on the CPR to Lake Wanapitei. Logs were driven down the Wanapitei River into the French River and then into Georgian Bay and finally towed to the Byng Inlet mills.

[15] Logging in Algonquin Park was a subject of much contention in the 1950s and 1960s. The Sheppard timber limits and many others had been licensed by the provincial government before the park was established in 1894. These licences were to have expired in 1930, after which no new ones were to have been issued; but future governments did issue licences for logging hardwood, thus precipitating a row with environmentalists and nature lovers.

[16] The Sheppard and Gill Lumber Co. was capitalized at $150,000, each of the three partners holding 500 of the 1,500 shares authorized. The company started off with a 2 1/2-acre lumber and coal yard on Drayton Avenue, Toronto; later it established yards in Richmond Hill and Bowmanville. Lumber was originally purchased from Georgian Bay Lumber, but when the mill at Waubaushene closed, Sheppard and Gill operated their own mill at Sultan, Ontario, on the CPR mainline, forty miles east of Chapleau. In 1951 the company's capital was increased to $650,000 with the issue of 500,000 one-dollar common shares. The Sheppards and Ernest Gill sold their interests in 1959. The company was dissolved in 1974.

CHAPTER FOURTEEN

[1] Ontario, *Sessional Papers, 1902*, Annual Report of the Director of Forestry for the Year 1900-1901.

[2] NAC, RG 10, vol. 2217, file 43,168-1, reel C-11,118, George S. Chitty to the Secretary, Department of Indian Affairs, 7 November, 1898.

[3] Ibid.

[4] NAC, reel C-11,118, Thomas Walton to the Superintendent General of Indian Affairs, 16 December 1888.

[5] NAC RG 10, vol. 2218, file 43,168-2A, reel C-11,1182, D.C. Scott to the Deputy Superintendent General of Indian Affairs, 14 August 1908.

[6] The "Robinson" Treaty, *Treaties and Surrenders*, 1, 149-52.

[7] NAC, RG 10, vol. 2,217, file 43168-7l, Walton to the Superintendent General of Indian Affairs, 14 December 1888.

[8] Ibid., Walton to the Superintendent General of Indian Affairs, [9] October 1886.

[10] There was no minister of Indian affairs then. The minister of the interior, acting also as superintendent general of Indian affairs, was responsible to Parliament for Indian affairs.

[11] NAC, RG 10, vol. 2,217, file 43168-71, Walton to to the Superintendemt General of Indian Affairs, 14 December 1888.

[12] Ibid., telegram, Harry Symons to S.R. Hesson, MP, 14 June 1887.

[13] Ibid., Walton to the Superintendent General of Indian Affairs, 14 December 1888.

[14] Ibid.

[15] Ibid.

[16] Ibid., L. Vankoughnet to the Hon. T.M. Daly, 1 March 1893.

[17] Ibid.

[18] Ibid., Michel Eagle Dokis to the Deputy Superintendent General of Indian Affairs, 10 July 1893.

[19] Ottawa *Evening Journal*, 27 April 1893.

[20] Ibid. Draft of a letter from Walton to Dokis band members, July 1893.

[21] Ibid., Walton to Vankoughnet, 21 August 1893.

[22] Ibid.

[23] Ibid.

[24] Ibid.

[25] Ibid.

[26] Ibid.

[27] Ibid.

[28] Ibid., Dokis to the Department of Indian Affairs, 25 May 1899.

[29] Ibid., Dokis to the Department of Indian Affairs, 6 April 1899.

[30] Timber regulation quoted in a memorandum from the secretary of the Department of Indian Affairs to the superintendent general, 7 March 1899.

[31] Ibid.

[32] Ibid., E.D. McLean to Dokis, 16 June 1899.

[33] Ibid., Dokis to the Department of Indian Affairs, 25 May 1899.

[34] Canada, *Debates*, 4 May 1908, 7807.

[35] NAC, RG 10, vol. 2219, file 43,168-6, reel 11182, George P. Cockburn to Frank Pedley, 6 February 1905.

[36] Ibid., Deputy Superintendent General of Indian Affairs to Cockburn, 3 March 1905.

[37] Ibid., Cockburn to Frank Pedley, 23 March 1905.

[38] Ibid., W.A. Orr to Assistant Secretary, Department of Indian Affairs, 5 November 1906.

[39] Ibid., Pedley to the Hon. Frank Oliver, 29 November 1906.

[40] Ibid., W.A. Orr to Deputy Superintendent General, 27 February 1907.

[41] Ibid., Pedley to Cockburn, 2 March 1907.

[42] Ibid., Pedley to Cockburn, 5 March 1907.

[43] Ibid.

[44] Ibid., telegram, Cockburn to Pedley, 8 January 1908.

[45] Ibid., affidavit signed by Chief Alex Dokis, 11 September 1908.

[46] Ibid., Chitty to Deputy Superintendent General, 12 January 1909.

[47] NAC, RG 10, vol. 2219, file 43168-6, reel C-11182, W.J. Sheppard to J.D. McLean, 28 June 1910.

[48] Ibid., Thomas A. Duff to Duncan C. Scott, 22 August 1918.

[49] Ibid., H.J. Bury to the Deputy Minister of Indian Affairs, 4 October 1918.

[50] Ibid., Bury to the Deputy Minister of Indian Affairs, 3 May 1919.

CHAPTER FIFTEEN

[1] *Canada Lumberman*, 1 January 1921.

[2] Canada, *Statutes*, 1917, 7-8 George V, c. 55, An Act to Authorize the Levying of a War Tax upon Certain Incomes.

[3] NAC, RG 95, vol. 1, Tilley, Johnson, Thomson, and Parmenter to the Undersecretary of State, 12 July 1921.

[4] When Ellen Sheppard died in 1928, she left Maud Skene $6,000 from an estate valued at slightly more than $100,000.

[5] Ellen Sheppard bequeathed $1,000 to the United church in Coldwater, distributed among the Board of Trustees, the Ladies' Aid, and the Women's Missionary Society.

[6] The company's charter, its annual returns, some financial statements, company by-laws, and relevant correspondence can be found in the NAC, Corporation Branch Records, RG 95, vol. 1.

APPENDIX A

Statement of Loss and Gain
Georgian Bay Mill Books, 30 April 1872

Waubaushene Mill

Cost of 11,842,748 ft Logs bot of Georgian Bay Lumber Co. at $5.00 per M.	$59,212.39	
Cost of towing, shipping, manufacturing, construction, teaming, building docks, houses, storehouses, cribs, repairing mill, new machinery from Feby 1, 1871 to May 1, 1872	74,797.19	$134,009.58
Lumber sold to amount of	88,403.04	
On hand, per inventory	32,602.72	121,005.76
Net loss on operations		13,003.82

Severn Mill

Cost of 12,427,206 logs bot of Georgian Bay Lumber Co, at $5.00 per M.	62,136.03	
Cost of manufacturing, towing, shipping, construction of buildings, docks, tramways, houses, repairing mill up to May 1, 1872	45,456.53	107,592.56
Lumber sold to amount of	92,769.22	
On hand, per inventory	31,321.50	124,090.72
Net gain on operations		16,498.16

Sturgeon Bay Mill

Cost of 1,164,010 ft logs bot of Georgian Bay Lumber Co. at $5.00 per M.	5,820.05	
Cost of manufacturing, towing, shipping	2,894.43	8,714.48
Lumber sold to amount of	7,520.04	
On hand, per inventory	6,120.00	13,640.04
Net gain on operations		4,925.56

Waubaushene Store

Goods on hand Feby 1, 1871	3,500.00	
Goods purchased, freight since Feby 1871 until May 1 1872	47,133.30	50,633.30
Goods sold to amount of	42,378.04	
On hand, per inventory	11,183.84	53,561.88
Net gain		2,928.58

Severn Store

Goods on hand Feby 1, 1871	2,000.00	
Goods purchased, freight since Feby 1871 to May 1, 1872	23,757.08	25,757.08
Goods sold to amount of	20,402.84	
On hand, per inventory	6,346.33	26,749.17
Net gain		992.09

Sturgeon Bay Store

Goods purchased from October 1871 up to		
December 30, 1871	677.80	
Goods sold to amount of	648.29	
Net loss		29.51

Thos. C. Street

Net gain season's trading		4,619.41
Queen of the North		
Net gain on season's trading		747.44
Kcnosha		
Net gain on season' trading		3,411.43
Dauntless		
Net gain on season's trading		358.39
Tug **Lilly Kerr**		
Net loss on season's work		105.30
Tug **Mittie Grew**		
Net loss on season's work		870.81
Amount of P. Milne's acct.	417.14	
Amount realized from Milne's acct.	337.38	
Net loss Milne's acct.		79.76

Recapitulation

Waubaushene Mill	13,003.82	
Severn Mill		16,498.16
Sturgeon Bay Mill		4,925.56
Waubaushcne Store		2,928.58
Severn Store		992.09
Sturgeon Bay Store	29.51	
Thos. C. Steet		4,619.41
Queen of the North		747.44
Kenosha		3,411.43
Dauntless		358.39
Tug **Lilly Kerr**	105.30	
Tug **Mittie Grew**	870.81	
P. Milne's acct.	79.76	
Net gain	20,391.86	
	34,481.06	34,481.06

APPENDIX B

Canadian Assets of William Earl Dodge, May 1883

1	Shares in the capital stock of Georgian Bay Lumber Co.	$199,000.00
2	Balance of account due by Georgian Bay Lumber Co.	231,009.84
3	Shares in the capital stock of Collingwood Lumber Co.	132,000.00
4	Balance of account due by Collingwood Lumber Co.	39,733.76
5	Shares of the capital stock of Maganettewan Lumber Co.	149,000.00
6	Balance of account due by Maganettewan Lumber Co.	103,700.63
7	Shares in the capital stock of Severn Driving and Boom Co.	1,200.00
8	Shares in the capital stock of Maganettewan Driving and Boom Co.	1,200.00
9	Balance of account due by Parry Sound Lumber Co.	21,666.66
10	Balance of account due by Longford Lumber Co.	45,000.00
	Total	$923,510.89

APPENDIX C

Properties of Georgian Bay Lumber Co., December 1873

This statement shows the cost of the various properties of the Georgian Bay Lumber Co. of Ontario and the additional value placed on them to represent the increase of market value and also to make up amount to represent the captial stock, as of December 1873.

	Costs	Actual Costs
Severn Mill		
Original purchase money	$25,000.00	
Construction to Nov.1, 1872	28,718.77	
Construction from Nov.1872 to Dec. 1, 1873	7,528.36	
700,000 feet of lumber at $6.50 per 1,000		
feet used in construction	4,550.00	
Actual cost		$65,797.13
Waubaushene Mill		
Original purchase money	35,531.24	
Construction to Nov. 1, 1872	42,853.36	
Construction from Nov. 1872 to Dec. 1, 1873	15,472.00	
1,100,000 feet of lumber at $6.50 per 1,000		
feet used in construction	7,150.00	
Actual cost		101,006.60
Christie limits: 70 1/4 square miles	36,714.00	
Laramy limits: 105 1/16 square miles	42,000.00	
Hall limits: 24 square miles	4,000.00	
Cook limits: 303 1/2 square miles	182,400.00	
Matchedash limits: 28 3/4 square miles	2,631.31	
Kennedy limits: 10 square miles	1,531.00	
Carden and Laxton limits: 37 square miles	5,845.30	
Actual Cost: 578 9/16 square miles		275,121.61
Deeded Lands		
Bought from A.R. Christie: 11,350 acres	45,400.00	
Bought from W. Hall: 4,500 acres	9,000.00	
Bought from Laramy: 850 acres	3,400.00	
Bought from various parties: 1,300 acres	10,949.59	
Sold to A. R. Christie: 1,446 acres	(5,864.00)	
Sold various parties	(479.09)	
Actual cost: 16,554 acres		62,406.50
Timber purchased		
From A.R.Christie: 2,962 acres	8,856.00	
From W. Hall: 1,000 acres	2,000.00	
From various parties	8,298.26	
Sold various parties	(1,013.24)	
Actual Cost: 13,413 acres		18,171.02

Orillia property		4,000.00

Vessel property
Original puchase money and permanent repairs

T.C. Street	16,000.00	
Queen of the North	13,000.00	
Kenosha	9,000.00	
Dauntless	1,500.00	
Lilly Kerr	6,000.00	
Prince Alfred	3,000.00	
Less Kenosha cost	(9,000.00)	
Actual cost: vessels		39,500.00

Personal property

On Hand Waubaushene and Orillia	35,560.50

Total cost: all properties	$601,563.45

Capital stock (made up as follows)

Cook Limits	$184,200.00	
Matchedash Limits	14,375.00	
Christie Limits	36,000.00	
Laramy Limits	42,000.00	
Hall Limits	4,000.00	
Kennedy Limits	1,000.00	
Carden and Laxton Limits	1,000.00	
Deeded Land	64,509.24	
Purchased Timber	5,000.00	
Severn Mill	65,797.13	
Waubaushene Mill	101,475.88	
Orillia Property	4,382.75	
Vessel Property	40,700.00	
Personal Property	35,560.00	
Total		$600,000.00

APPENDIX D

Lumber Companies of A.G.P. Dodge

Parry Sound Lumber Company 1872-76

Capital stock	$300,000
Number of shares	3,000
Share value	$100

Shareholders	**Shares**
A.G.P. Dodge	2,800
John Gilchrist	Distribution of remaining
William J. Hunt	shares not known
D'Alton McCarthy	
J.C. Miller	
David Crawford White	

President: A.G.P. Dodge
Secretary: D'Alton McCarthy
Treasurer: William J. Hunt

Maganettewan Lumber Co., 1871-72 (Dominion charter)

Capital stock	$210,000
Number of shares	2,100
Share value	$100

Shareholders	**Shares**
A.G.P. Dodge	1,510
George William Allen	10
Charles Henry Dill	225
Charles Hebbard	20
George Kempt	10
William D. Kintzing	20
Titus B. Meigs	50
Levi Miller	225
Angus Morrison	10
Strachan Napier Robinson	10
Samuel Schofield	10

President: A.G.P. Dodge
Secretary: Not known
Treasurer: Not known

Maganettawan Lumber Co. of Ontario, 1872-75

Capital stock	$700,000
Number of shares	7,000
Share value	$100

Shareholders	**Shares**
A.G.P. Dodge	
Samuel White Bernard	
Eli Clinton Clarke	
Eli Clinton Clarke Jr	
Charles Henry Dill	
William D. Kintzing	Share distribution not known

Levi Miller
Harvey Mixer
D'Alton McCarthy
Alanson Sumner Page
Douglas Leland White

Officers: not known

Georgian Bay Lumber Co., 1871 (Dominion charter)

Capital stock	$229,500
Number of shares	2,295
Share value	$100

Shareholders	Shares
A.G.P. Dodge	2,000
Alexander Christie	10
William Kerr	10
D'Alton McCarthy	10
William J. Macaulay	250
John Beverley Robinson	5
Daniel Sprague	10

President: A.G.P. Dodge
Secretary: D'Alton McCarthy
Treasurer: Daniel Sprague

Georgian Bay Lumber Co. of Ontario,1872

Capital stock	$1,000,000
Number of shares	10,000
Share value	$100

Shareholders	Shares
A.G.P. Dodge	9,830
Alexander Christie	10
D'Alton McCarthy	10
Harvey Mixer	100

President: A.G.P. Dodge
Secretary: D'Alton McCarthy
Treasurer: D'Alton McCarthy

Longford Lumber Co.

Unincorporated co-partnership with John
   Thomson

Muskoka Mills Lumber Co.

Unincorporated co-partnership with J.C.
   Hughson

APPENDIX E

Lumber Companies of William Earl Dodge

Maganettewan Lumber Co., 1876-82

Capital stock	$300,000
Number of shares	3,000
Share value	$100

1876-82

Shareholders	Shares
T.W. Buck	25
Charles Dill	5
George E. Dodge	25
Norman Dodge	25
W.E. Dodge	2,920

President: W.E. Dodge
Secretary: T.W. Buck
Treasurer: T.W. Buck

1882-83

Shareholders	Shares
A.M. Dodge	1,500
C.H. Dodge	5
W.E. Dodge	1,485
J. Peckham	5
J. Scott	5

President: W.E. Dodge
Secretary: J. Scott
Treasurer: J. Scott

Parry Sound Lumber Co.,1876-77

Capital stock	$300,000
Number of shares	3,000
Share value	$100

1876

Shareholders	Shares
T.W. Buck	5
A.G.P. Dodge	10
W.E. Dodge	2,800
D'Alton McCarthy	10
J.C. Miller	165
Harvey Mixer	10
Harvey Mixer	10

President: W.E. Dodge
Secretary: D'Alton McCarthy
Treasurer: T.W. Buck

1877

Shareholders	Shares
T.W. Buck	5
Alex Campbell	10
A.G.P. Dodge	10
D'Alton McCarthy	10
Thomas McCrachen	100
J.C. Miller	2,855

President: J.C. Miller
Secretary: not known
Treasurer: not known

Muskoka Mills Lumber Co., 1875

Capital stock	$300,000
Number of shares	3,000
Share value	$100

Shareholders	Shares
T.W. Buck	60
A.H Campbell	451
W.E. Dodge	720
J.C. Hughson	450
J.S. Huntoon	300
T.B. Meigs	720
N.H. Salisbury	299

President: A.H. Campbell
Secretary treasurer: T.W. Buck
Superintendent: J.S. Huntoon

Georgian Bay Lumber Co., 1876-81

Capital stock	$600,000
Number of shares	6,000
Share value	$100

Shareholders	Shares
T.W. Buck	5
George E. Dodge	5
Norman Dodge	5
W.E. Dodge	3,000
Dodge, Meigs & Co.	2,980
T.B. Meigs	5

President: W.E. Dodge
Secretary: T.W. Buck
Treasurer: T.W. Buck

Georgian Bay Lumber Co., 1881-83

Capital stock	$400,000
Number of shares	4,000
Share value	$100

Shareholders	Shares
A.M. Dodge	2,005
C.H. Dodge	5
W.E. Dodge	1,980
J. Peckham	5
J. Scott	5

President: W.E. Dodge
Secretary treasurer: J. Scott
Superintendent: J. Peckham

Collingwood Lumber Co., 1878-83

Capital stock	$150,000
Number of shares	1,500
Share value	$100

1878

Shareholders	Shares
T.W. Buck	25
David Cooper	5
A.M. Dodge	375
W.E. Dodge	1,090
J. Scott	5

President: W.E. Dodge
Secretary treasurer: J. Scott
Superintendent: David Cooper

1882-83

Shareholders	Shares
A.M. Dodge	175
C.H. Dodge	5
W.E. Dodge	1,310
J. Peckham	5
J. Scott	5

President: W.E. Dodge
Secretary treasurer: J.Scott
Superintendent: J.Peckham

Longford Lumber Co., 1876

Capital stock	$300,000
Number of shares	3,000
Share value	$100

Shareholders	Shares
T.W. Buck	50

W.E. Dodge	1,400
A.G. Longford	20
T.B. Meigs	50
John Thomson	1,400

President: John Thomson
Secretary treasurer: John Thomson
Superintendent: Alex Thurburn

Severn Driving and Boom Co., 1882

Capital stock	$50,000
Number of shares	500
Share value	$100

Shareholders	Shares
A.M. Dodge	160
C.H. Dodge	80
W.E. Dodge	200
H.L. Lovering	20
J.S. Peckham	20
J. Scott	20

President: W.E. Dodge
Secretary treasurer: J. Scott

Maganettewan Driving and Boom Co.,1883

Capital stock	$ 50,000
Number of shares	500
Share value	100

Shareholders	Shares
A.M. Dodge	180
C.H. Dodge	60
W.E. Dodge (estate)	200
H.L. Lovering	20
J.S. Peckham	20
J. Scott	20

President: A.M. Dodge
Secretary treasurer: J. Scott

APPENDIX F

Lumber Companies of Arthur Murray Dodge

Georgian Bay Consolidated Lumber Co., 1883-92

Capital stock	$1,000,000
Number of shares	10,000
Share value	$100

Shareholders	Shares
A.M. Dodge	3,770
C.H. Dodge	1,250
W.E. Dodge (estate)	4,965
W.E. Dodge Jr	5
J. Peckham	5
J. Scott	5

President: A.M. Dodge
Secretary treasurer: J. Scott
Superintendent: J. Peckham

Georgian Bay Lumber Co., 1893

Capital stock	$200,000
Number of shares	2,000
Share value	$100

Shareholders	Shares
Arthur M. Dodge	1,075
J.W. Hartman	25
J.L. Lovering	225
W.H.F. Russell	75
J. Scott	225
Thomas H. Sheppard	110
W.J. Sheppard	225
C.P. Stocking	40

President: A.M. Dodge
Secretary: C.P. Stocking
Treasurer: J. Scott

Dodge and Bliss Box Co., 1886

Capital stock	$50,000
Number of shares	500
Share value	$100

Shareholders	Shares
Delos Bliss	105
A.M. Dodge	105
Jefferson Hunt	5
J.S. Peckham	5
J. Scott	5

President: A.M. Dodge
Secretary treasurer: J.Scott

Dominion Mercantile Co., 188-93

Capital stock	$50,000
Number of shares	500
Share value	$100

Shareholders	Shares
A.M. Dodge	173
C.H. Dodge	57
W.E. Dodge Jr	5
J.W. Hartman	5
W.H.F. Russell	5
J. Scott	5

President: A.M. Dodge
Secretary treasurer: W.H.F. Russell

APPENDIX G

Lumber Companies of W.J. Sheppard

Sheppard Lumber Co., 1899-1924

Capital stock	$85,000
Number of shares	850
Share value	$100

Shareholders (1906)	Shares
Frederick M. Gray	282
Annie Gray	2
C.H.Sheppard	282
E.F. Sheppard	50
Leigh B. Sheppard	50
Robert S. Sheppard	50
W.J. Sheppard	182

President: W.J. Sheppard
Secretary treasurer: C.H. Sheppard

Muskoka Lakes Lumber Co. 1909-1924

Capital stock	$30,000
Number of shares	300
Share value	$100

Shareholders	Shares
E. Breech	1
G.A. Miller	1
C.H. Sheppard	72
E.F. Sheppard	1
W.J. Sheppard	225

President: W.J. Sheppard
Secretary treasurer: C.H. Sheppard

Sheppard and Dunn Lumber Co., 1909-1929

Capital stock	$42,000
Number of shares	420
Share value	$100

Shareholders	Shares
John T. Dunn	
Annie M. Gray	
C.H. Sheppard	
H.L. Sheppard	Distribution unknown
L.B. Sheppard	
R.S. Sheppard	
W.J. Sheppard	
W.J. Sheppard Jr	

President: W.J. Sheppard
Secretary treasurer: C.H. Sheppard

Dunn Lumber Co., 1909-1924

Capital stock	$15,000
Number of shares	150
Share value	$100

Shareholders	Shares
J.T. Dunn	50
Fred M. Gray	25
W.H.F. Russell	25
C.H. Sheppard	25
W I Sheppard	25

President: W.J. Sheppard
Secretary treasurer: C.H. Sheppard

Metagama Lumber Co., 1899-1904

Capital stock	$100,000
Number of shares	1,000
Share value	$100

Shareholders	Shares
O. Arnold	100
W.R. Beatty	100
William Irwin	100
James Ludgate	100
G. McCormick	100
Angus McLeod	100
John J. McNeil	100
W.C. Mahaffy	100
Thomas H. Sheppard	100
W.J. Sheppard	100

President: George McCormick
Secretary treasurer: James White

Georgian Bay Lumber Co., 1926

Capital stock	$11,725
Number of shares	11,725
Share value	$1

Shareholders	Shares
Marshall Dodge	6
W.H.F. Russell	25
C.H. Sheppard	25
E.M. Sheppard	214
W.J. Sheppard	11,315
C.P. Stocking	40

President: W.J. Sheppard
Secretary treasurer: C.H. Sheppard

APPENDIX H

Lumber Companies of C.H. Sheppard

Georgian Bay Lumber Co., 1936-1942

Capital stock	$11,725
Number of shares	11,725
Share value	$1

Shareholders	**Shares**
R.E. Breech	1
Marshall Dodge	6
C.H. Sheppard	11,718

Vice-President: C.H. Sheppard
Secretary treasurer: R.E. Breech

Sheppard and Dunn Lumber Co., 1933-1965

Capital stock	$10,000
Number of shares	100
Share value	$100

Shareholders	**Shares**
R.E. Breech	1
J.T. Dunn	50
C.H. Sheppard	48
L.B. Sheppard	1

President: Jack Dunn
Secretary treasurer: Ella Breech

Parry Sound Public Library
  Records and photographs
Simcoe County Archives (Minesing)
  Records and photographs
Simcoe County Land Registry Office (Barrie)
  Nottawasaga township, conc. 9
  Orillia township, conc. 2
  Rama township, conc. 4
  Tay township, concs 9, 10, 11, 12, 13
Vermilion County Courthouse (Danville)
  Marriage Records and Wills
Waubaushene Public Library
  Records
West Parry Sound District Museum (Parry Sound)
  Records

STATUTES AND OTHER GOVERNMENT PUBLICATIONS

Canada

Canada *Gazette*
House of Commons, *Debates*, 1908
Statutes, 1869, 32-33 Vic., c. 12, The Canada Joint Stock Companies Act
  – 1869, 32-33 Vic., c. 73, An Act to Naturalize Eli Clinton Clark
  – 1872, 35 Vic., c. 118, An Act to Naturalize Anson Greene Phelps Dodge
  – 1894, 57-58 Vic., c. 107, An Act to Incorporate the French River Boom Company (Limited)
  – 1904, 4 Edward VII, c. 78, An Act Respecting the French River Boom Company Limited
  – 1917, 7-8 George V, c. 28, An Act to Authorize the Levying of a War Tax upon Certain Incomes
  – 1919, 9-10 George V, c. 55, An Act to Amend The Income War Tax Act, 1917

Ontario

Ontario *Gazette*
Sessional Papers, 1868-1920
  – Reports of the Commissioner of Crown Lands, 1868-1904
  – Reports of the Minister of Lands, Forests and Mines, 1905-1920
Statutes, 1869, 32 Vic., c. 30, An Act to Incorporate the Toronto, Simcoe and Muskoka Junction Railway
  – 1871, 34 Vic., c. 71, An Act to Incorporate the Georgian Bay Lumber Association
  – 1872, 35 Vic., c. 47, An Act to Incorporate the Parry Sound Lumber Company
  – 1872, 35 Vic., c 99, An Act to Incorporate the Georgian Bay Lumber Company of Ontario
  – 1872, 35 Vic., c. 100, An Act to Incorporate the Maganettewan Lumber Company of Ontario
  – 1873, 36 Vic., c. 74, An Act to Incorporate "The Lake Simcoe Junction Railway"
  – 1874, 37 Vic., c. 35, An Act Respecting the Incorporation of Joint Stock Companies by Letters Patent
  – 1881, 44 Vic., c. 11, An Act for Protecting the Public Interest in Rivers, Streams and Creeks
  – 1881, 44 Vic., c. 19, An Act for the Incorporation by Letters Patent and Regulation of Timber Slide Companies
  – 1890, 53 Vic., c. 7, An Act Respecting the Culling and Measurement of Saw-Logs Cut upon Crown Lands
  – 1898, 61 Vic., c. 9, An Act Respecting the Manufacture of Pine Cut on the Crown Domain

311

NEWSPAPERS

Barrie, *Gazette*
Barrie, *Northern Advance*
Bracebridge, *Northern Advocate*
Coldwater and Waubaushene, *Tribune and Investigator*
Danville, *Daily News*
Danville, *Monetary Times*
Hamilton, *Times*
Midland, *Free Press*
Newmarket, *Courier*
Newmarket, *Era*
Newmarket, *Express-Herald*
New York, *Times*
Orillia, *Northern Light*
Orillia, *Packet*
Orillia, *Packet and Times*
Orillia, *Times*
Orillia, *Times and Expositor*
Toronto, *Canada Lumberman and Woodworker*
Toronto, *Globe*
Toronto, *Mail*
Toronto, *Monetary Times*

BOOKS AND ARTICLES

Bruchey, Stuart, and Eleanor Bruchey (eds). *The Lumber Industry*. New York: Arno Press 1972

Brunton, Sam. *Notes and Sketches on the History of Parry Sound*. Parry Sound: Parry Sound Public Library 1969

Buggey, Susan, and John J. Stewart. "Lakehurst and Beechcroft: Roches Point, Ontario, Canada." *Journal of Garden History*. 1, no. 2 (1981): 147-66

Canadian Parliamentary Guide, 1868-1920

Cleland, Robert G., *A History of Phelps Dodge, 1834-1950* New York: Alfred A. Knopf 1950.

Cross, Michael S., "The Lumber Community of Upper Canada." *Ontario History*. (52, no. 4 (1960):) 213-34

Defebaugh, James Elliott. *History of the Lumber Industry of America*. 2 vols. Chicago 1906

Dodge, D. Stuart. *Memorials of William E. Dodge*. New York: Anson D.F. Randolph 1887

Dodge, Phyllis B. *Tales of the Phelps-Dodge Family: A Chronicle of Five Generations*. New York: New York Historical Society 1987

Drury, E.C. *All for a Beaver Hat: A History of Early Simcoe County*. Toronto 1959

Dunnan, Nancy, and Pauline Dodge. *The Jewells and the Dodges: An American Saga 1635-1828*. New York 1987

Easterbrook, W.T., and Hugh G.J. Aitken. *Canadian Economic History*. Toronto: Macmillan 1975

French, Gary E. "William Basil Hamilton, Penetanguishene Fur Trader and Collingwood Pioneer." *East Georgian Bay Historical Journal*. 111, (1983): 16-42

Gazetteer and Directory of the County of Simcoe. Elmvale. Ontario: East Georgian Bay Historical Foundation 1985

Hamilton, J.C. *The Georgian Bay: An Account of its Position, Inhabitants, Mineral Interests, Fish, Timber, and Other Resources*. Toronto 1893

Hunter, Andrew F. *A History of Simcoe County*. Barrie: Historical Committee of Simcoe County 1948

Indian Treaties and Surrenders. Toronto: Coles Canadiana Series 1971

Jackson, J.B. *Lumberman's Timber Mark Guide*. Toronto 1874

Jones, Owen. "The Mills at the Mouth of the Musquash River." *Cognashene Cottager*.(Summer 1980)

Josephson, Matthew. *The Robber Barons*. New York: Harcourt, Brace and World 1934

Kauffmann, Carl. *Logging Days in Blind River*. Sault St Marie: Sault Star Commercial Printing 1970

Kirkwood, Alexander, and J.J. Murray. *The Undeveloped Lands in Northern and Western Ontario . . . Collected and Compiled from Reports of Surveyors, Crown Land Agents, and Others, with the Sanction of the Commissioner of Crown Lands.* Toronto 1878.

Knight, Rolf. *Work Camps and Company Towns in Canada and the U.S.: An Annotated Bibliography*. Vancouver: New Star Books 1975

Lambert, Richard S. and Paul Pross. *Renewing Nature's Wealth A Centennial History.* Toronto. Department of Lands and Forests 1967

Leitch, Adelaide. *The Visible Past*. Toronto: Ryerson 1967

Lower, A.R.M. *The North American Assault on the Canadian Forest*. Toronto: Ryerson 1938
– *Settlement and the Forest Frontier in Eastern Canada*. Toronto 1936
– "The Trade in Square Timber." University of Toronto, *Studies* History and Economics, Contributions to Canadian Economics. Toronto 1933, 40-61.

Lowitt, Richard. *A Merchant Prince of the Nineteenth Century, William E. Dodge*. New York: Columbia University Press 1954

Macfie, John. *Now and Then Footnotes to Parry Sound History.* Parry Sound: Beacon Publishing 1983
– *Parry Sound Logging Days*. Erin, Ontario: Boston Mills Press 1987

Mackay, Martin. *Over the Hills to Georgian Bay*. Erin, Ontario: Boston Mills Press 1981

MacKay, Donald. *The Lumberjacks*. Toronto: McGraw-Hill Ryerson 1978

Martyn, Carlos. *William E. Dodge: The Christian Merchant*. New York: Funk and Wagnalls 1890

McMurray, Thomas. *The Free Grant Lands of Canada, from Practical Experience of Bush Farming in the Free Grant Districts of Muskoka and Parry Sound*. Bracebridge 1871

Michell, H. "Profit-Sharing and Producers' Co-operation in Canada." *Bulletin of the Department of History and Poliical and Economic Science in Queen's University*. Kingston 1918

Morris, Alexander. *The Treaties of Canada with the Indians*. Toronto: Belford and Clarke 1880

Murray, Florence B. *Muskoka and Haliburton* 1615-1875. Toronto: The Champlain Society 1963

Nelles, V.H. *The Politics of Development: Forests, Mines & Hydro-Electric Power in Ontario*, 1849-1941. Toronto: Macmillan 1974

Obituary Record of Graduates of Yale University. New Haven: Yale University Press 1897

One Hundred Years of the Coldwater United Church. Coldwater 1966

Patstone, Arthur J. *A Short History of St. Paul's Church, Newmarket*. Toronto: Church of England (undated)

Pearson, G. C. *The Past and Present of Vermilion County, Illinois*. Danville 1908

Price, Eugenia. *The Beloved Invader*. New York: Avon Books 1965
– *St. Simons Memoir*. New York: J.B. Lippincott 1978

Radforth, Ian Walter. *Bushworkers and Bosses: Logging in Northern Ontario, 1900-1980*. Toronto: University of Toronto Press 1987.

Rawson, Mabel. *The Story of Port Severn Yesterday, To-day and To-morrow*. Midland: Simcoe Press 1976

Simcoe County. *Illustrated Atlas*. Toronto 1970

Stevens, G.R. *Canadian National Railways: Sixty Years of Trial and Error, 1836-1896*. 2 vols. Toronto: Clarke, Irwin 1960

Trout, J.M., and Edward Trout. *The Railways of Canada for 1870-1.* Toronto: Monetary Times 1871

Whitton, Charlotte. *A Hundred Years A-Fellin'*. Braeside, Ontario: Gillies Brothers 1974

Index

320